T0132663

GENE PROFILES IN DRUG DESIGN

GENE PROFILES IN DRUG DESIGN

Edited by
BRETT A. LIDBURY
and
SURESH MAHALINGAM

CRC Press
Taylor & Francis Group
Boca Raton   London   New York

CRC Press is an imprint of the
Taylor & Francis Group, an **informa** business

CRC Press
Taylor & Francis Group
6000 Broken Sound Parkway NW, Suite 300
Boca Raton, FL 33487-2742

© 2009 by Taylor & Francis Group, LLC
CRC Press is an imprint of Taylor & Francis Group, an Informa business

International Standard Book Number-13: 978-0-8493-3733-8 (Hardcover)

Library of Congress Cataloging-in-Publication Data

Gene profiles in drug design / editors, Brett A. Lidbury and Suresh Mahalingam.
    p. ; cm.
  "A CRC title."
  Includes bibliographical references and index.
  ISBN 978-0-8493-3733-8 (hardcover : alk. paper) 1. Pharmacogenetics. 2. Drugs--Design. I. Lidbury, Brett A. II. Mahalingam, Suresh. III. Title.
  [DNLM: 1. Drug Design. 2. Gene Expression Profiling. 3. Gene Expression Regulation. 4. Gene Therapy--methods. 5. MicroRNAs--therapeutic use. QV 744 G326 2008]

RM301.3.G45.G458 2008
615'.7--dc22
                                                  2008013724

Visit the Taylor & Francis Web site at
http://www.taylorandfrancis.com

and the CRC Press Web site at
http://www.crcpress.com

Contents

Preface

Almost unbelievably, virtually every human gene is now known. This new "Post-Genomics" era promises that the vast amounts of information generated by genomics (and proteomics) will rapidly lead to better drugs and therapies. The irony of the genomics triumph, however, has been that it is easier to find and measure genes than to manipulate genetic pathways and then apply the results to human health. In the "Post-Genomics" era, the major challenge has shifted from identifying genes to explaining how genes work in concert and, importantly, how these genes cause disease or participate in disease processes. A quick and easy route to new drugs emanating from the genetic revolution appears to be out of reach.

The challenges can appear insurmountable; the technical and logistical problems are all new, and the technologies are generally untried. In spite of this, some are taking on these challenges and risks in the hopes of converting genomics data into products that will help diagnose, treat, and prevent disease. The research and technologies of this newly emerging field vary widely in design and implementation, but they all share a core principle in improving human health. Gene profiles generated from genomics data can be used not only to test and improve existing drugs, but also to create new drugs. The emerging new technologies promise to improve health, lower costs, decrease side effects, and reduce development time compared with more traditional methods.

This book combines views and insights from leaders in the pharmaceutical industry and academic research, as well as analysis of an enhanced genetic future for health and medicine from a clinical and bioethical perspective. Therefore, *Gene Profiles in Drug Design* is a book that is intended for anyone who wishes to explore scientific and technological processes associated with gene-profile-based drug design as well as the clinical and ethical implications of advances in this field.

C. Allen Black, J.D., Ph.D.
Pittsburgh, Pennsylvania

Brett A. Lidbury, Ph.D.
Virus and Inflammation Research Group, Centre for Biomolecular and
Chemical Sciences
University of Canberra, Australia

Acknowledgment

The editors wish to thank Dr. C. Allen Black for the inception of the Gene Profiles in Drug Design Project and for the early insights and ideas that contributed to the content and format of the final volume.

Contributors

Eric A. G. Blomme
Global Pharmaceutical Research and
  Development, Abbott
Abbott Park, Illinois, USA

Lexie Brans
Faculty of Health
University of Canberra
Canberra, Australia

Mark T. Heise
The Carolina Vaccine Institute,
  Department of Genetics and
  Department of Microbiology and
  Immunology
University of North Carolina
Chapel Hill, North Carolina, USA

Jasjit Johal
Virus and Inflammation Research
  Group, Centre for Biomolecular and
  Chemical Sciences
University of Canberra
Canberra, Australia

Brian H. Johnston
Somagenics, Inc.
Santa Cruz, California, USA

Les P. Jones
NIAID Center of Excellence for
  Influenza Research and Surveillance,
  Department of Infectious Diseases,
  Animal Health Research Center
University of Georgia
Athens, Georgia, USA

Sergei A. Kazakov
Somagenics, Inc.
Santa Cruz, California, USA

Brett A. Lidbury
Virus and Inflammation Research
  Group, Centre for Biomolecular and
  Chemical Sciences
University of Canberra
Canberra, Australia

Suresh Mahalingam
Virus and Inflammation Research
  Group, Centre for Biomolecular and
  Chemical Sciences
University of Canberra
Canberra, Australia

Cristina M. Musso
Virus and Inflammation Research
  Group, Centre for Biomolecular and
  Chemical Sciences
University of Canberra
Canberra, Australia

Artem S. Rogovskyy
NIAID Center of Excellence for
  Influenza Research and Surveillance,
  Department of Infectious Diseases,
  Animal Health Research Center
University of Georgia
Athens, Georgia, USA

Nestor E. Rulli
Virus and Inflammation Research
  Group, Centre for Biomolecular and
  Chemical Sciences
University of Canberra
Canberra, Australia

Dimitri Semizarov
Global Pharmaceutical Research and
  Development, Abbott
Abbott Park, Illinois, USA

Julian W. Tang
Department of Microbiology
The Chinese University of Hong Kong,
  Prince of Wales Hospital
Shatin, New Territories, Hong Kong,
  China

Beverly A. Teicher
Genzyme Corporation
Framingham, Massachusetts, USA

Stephen Mark Tompkins
NIAID Center of Excellence for
  Influenza Research and Surveillance,
  Department of Infectious Diseases,
  Animal Health Research Center
University of Georgia
Athens, Georgia, USA

Ralph A. Tripp
NIAID Center of Excellence for
  Influenza Research and Surveillance,
  Department of Infectious Diseases,
  Animal Health Research Center
University of Georgia
Athens, Georgia, USA

1 A Perspective on the Future Clinical Impact of Genetic Diagnosis and Gene-Based Drug Therapies for Patient Health

Julian W. Tang
Department of Microbiology,
The Chinese University of Hong Kong

CONTENTS

I remember first hearing about the human genome project [1, 2], whilst still a second year medical student at Christ's College, Cambridge in 1988. Cambridge had good cause to be interested and excited about the project, since James Watson and Francis Crick were working together at Cambridge before sharing the Nobel Prize in Physiology or Medicine for their discovery of DNA in 1962, along with Maurice Wilkins—also a Cambridge graduate. During later years, I followed the rivalry between the two human genome sequencing projects with great interest. One was a privately-funded team led by *Craig Venter* (Celera Genomics) and the other was a publicly-funded, international team, led by James Watson that included groups from the USA (National Institutes of Health) and UK (Welcome Trust). The UK-based efforts took place at the Sanger Centre (now the Sanger Institute), which was and still is, sited at Cambridge—closing a loop started by Watson and Crick, 50 years earlier.

1.1   INTRODUCTION

Since the initial completion of the human genome project (sequencing the entire >3 billion base pairs in the human genome) in 2001 [3, 4], much has been written about how these data can be used for the benefit of mankind. The (sometimes fierce) rivalry between the privately and publicly funded efforts often revolved around whether or not to allow certain human genes to be patented for commercial exploitation later. Today's increasingly stringent demands for maintaining patients' confidentiality have been driven partly by the increasing incidence of human immunodeficiency virus (HIV) infections that began in the late 1980s, as well as the rapid increase in the use of the Internet, which allows patient data to be transferred virtually instantaneously anywhere across the world. The need (or desire) to keep one's genetic information "secret" has now become a subject of intense debate, and the penalties for not doing so have become proportionately more severe. Other events, such as the controversy over tissue retention in the United Kingdom [5], have contributed to the need for explicit, informed consent for the use of confidential patient data or tissue, especially with the widespread inclusion of HIV and high-risk genetic diseases (masked as questions about family history) and related questions on insurance applications.

Hence, the completion of the human genome project and the use of gene-profiling to improve and individualize patient care may, ironically, have come at one of the most sensitive times in health care, where the rights of individuals to assert the ownership and privacy of their genetic information have never been greater.

1.2   SOME HUMAN GENE POLYMORPHISMS THAT MAY AFFECT CLINICAL OUTCOMES

There are many examples of where gene profiling may be beneficial to patient outcomes. However, some inherited gene polymorphisms have implications beyond the immediate medical problem, particularly if the results of their genetic tests are recorded in their medical notes, which may be accessible to insurance companies. This is a major concern of patients being tested for HIV and genetic disorders. Other gene polymorphisms are also significant, but mainly have implications for the ease of managing the patient's condition, rather than other, nonmedical reasons.

With regard to predicting future risks, in the field of cancer, two genes, BRCA1 and BRCA2, have been identified as a significant risk factor for the development of breast and ovarian cancer in women and prostate and bowel cancer in men [6, 7]. These studies showed that early genetic testing can lead to effective intervention to prolong the life of such individuals, though there maybe a severe price to pay (e.g., early, prophylactic mastectomy). The problems of maintaining confidentiality and obtaining insurance were also highlighted in these two studies. As with HIV testing, testing for the presence of gene polymorphisms that predispose to cancer have obvious implications for any long-term economic commitments, such as obtaining a mortgage and different types of insurance. This is because the individual may not live long enough to pay all the installments or premiums. Thus, some individuals prefer not to be tested and not to know.

Other examples of gene polymorphisms that predict future risks and with similar levels of life-altering significance include other inherited genes predisposing to colon cancer (familial adenomatous polyposis, FAP) [8] as well as Huntington's disease. Huntington's disease (HD) is an inherited autosomal dominant disorder of late-onset nervous system degeneration manifested by involuntary movements and altered personality and behavior. The problem is that it only becomes manifest during middle age, usually after having had children, when the HD gene may have already been passed on to the next generation. Hence, children of one parent with HD know that they have a 50% chance of inheriting the disease, and the decision about genetic testing may be difficult [9, 10].

Variant Creutzfeldt-Jakob disease (vCJD) is an interesting combination of a transmissible agent with a genetic predisposition for the development of disease. If the individual has been exposed to the vCJD agent and has become infected, there are certain genotypes that make them more or less susceptible to disease. Other conditions with a possible vertical (inheritable) genetic component and a horizontally transmissible factor, usually an infectious agent of some kind, include Epstein-Barr virus and nasopharyngeal carcinoma, human papillomavirus (HPV) and cervical cancer, hepatitis B and C and hepatocellular carcinoma, and Kaposi's sarcoma and HIV/human herpesvirus 8 (HHV-8).

There is now accumulating evidence that vCJD may be transmitted by blood transfusion, as well as by eating contaminated meat, and by cross-infection from contaminated surgical instruments and transplant organs [11–14]. This poses quite different implications for screening individuals for vCJD, as compared with inherited single-gene disorders, which are not transmissible from person to person. Since the lag time between exposure/infection and onset of vCJD disease can be decades [15], blood donations may come from volunteers who do not know that they are incubating the disease. As before with HIV and hepatitis C, blood tests developed to screen blood donors to protect the blood supply also identified many previously unknown and unsuspecting individuals with these infections, often with life-altering consequences [16]. A similar situation may arise again if an accurate vCJD test is developed. Previously, it was thought that a certain type of gene polymorphism in the naturally existing prion protein gene (PRNP coding for the normal prion protein PrP^C), 129M/M (M, the amino acid methionine), was a necessary requirement for the development of human vCJD after exposure to contaminated meat containing the abnormal, infectious form of the prion protein, PrP^{Sc}. This was because all the earlier cases of fatal human vCJD exhibited only this 129M/M genotype, i.e., none of these confirmed vCJD cases expressed a 129M/V or a 129V/V (V, the amino acid valine) PRNP genotype [17]. So, initially, it was thought that perhaps other individuals without this 129M/M genotype might be naturally immune to vCJD infection and disease, or perhaps, alternatively, they may have even longer incubation periods [14], which would change the future pattern of the epidemic significantly [15].

But, would such individuals still be able to carry the abnormal PrP^{Sc} in their blood and transmit it to others via blood transfusion, while they themselves were asymptomatic? In fact, two types of tests are important for determining the risk of acquiring vCJD. One is a test of infection with the abnormal prion protein, PrP^{Sc}. This will determine whether a person is carrying this protein in his or her system, and therefore whether it may lead to the disease in that individual and/or whether it may be transmitted to others, e.g., via contaminated surgical instruments or organ transplantation. The other test is a genetic one, to determine the polymorphism of the host prion protein gene, PRNP, i.e., 129M/M, 129M/V, or 129V/V. This may determine the susceptibility of that individual to the abnormal PrP^{Sc}. Thus, an individual carrying the abnormal PrP^{Sc} protein may be able to transmit this to others, as well as having a variable risk of developing vCJD, depending on the individual's PRNP gene polymorphism. Both of these presymptomatic tests for the presence of the prion protein PrP^{Sc} or the gene polymorphism predisposing to the development of vCJD may therefore be important for blood donors. If such tests become available, then it is conceivable that any individual scheduled for surgery may also require preoperative testing for PrP^{Sc} carriage so that any surgical instruments used can then be quarantined away, separately, and not used on otherwise "negative" individuals, to avoid cross-infection. This is because the sterilization procedures for prion-contaminated surgical instruments is not yet 100% reliable, and some surgical procedures simply cannot use disposable instruments. Unlike blood donation, which is voluntary, some operations are life-saving and cannot be avoided. So the issue of testing to protect other individuals from cross-contamination may become mandatory, as most health-care institutions cannot afford to dispose of high-quality steel surgical instruments after a single use [16].

The significance of PNRP genotypes became even more uncertain with the finding of a vCJD case who had a 129M/V genotype [17], suggesting that there may be a gradation of susceptibility and/or incubation periods for each of the possible genotypes, i.e., 129 M/M being more susceptible with a shorter incubation period than 129M/V and, similarly, for 129M/V and 129V/V. Hence, vCJD is a good example of how a genetic element may interact with a transmissible element, how the ethical implications of testing diagnosis may affect the lives of individuals, and how our understanding of the disease may change as new data arise.

Another inherited form of genetic polymorphism that can affect an individual patient's current management (as opposed to predicting future risk) is described in a recent study, which demonstrated that there is an advantage to testing for a genetic polymorphism that affects how warfarin is metabolized by an individual. Warfarin is an anticoagulant drug used in many disorders where individuals have a tendency to form life-threatening clots in their blood vessels. The drug reduces the risk of clotting in conditions that may predispose to this, such as procoagulant autoimmune disorders like antiphospholipid syndrome, the presence of a metal replacement heart valve, or an irregular heartbeat (e.g., atrial fibrillation). The drug is cleared (or metabolized) by the body in the liver, where a hepatic enzyme, the cytochrome P450 isozyme CYP2C9, breaks down the drug for excretion. The gene coding for CYP2C9 (and its associated variant genotypes), therefore, determines how quickly or slowly warfarin is cleared from the circulation. This has implications for how frequently and in what dose the warfarin should be taken by the patient. If not enough warfarin is taken, the patient may die or become severely disabled from life-threatening clots (e.g., ischemic stroke, myocardial infarction, or pulmonary embolism). If the warfarin levels are too high because they are not metabolized at the expected rate, then bleeding may arise and cause life-threatening or severe disease (e.g., hemorrhagic stroke or uncontrollable intra- and postoperative bleeding). This study showed that the correct warfarin dosing for an individual patient, based on his or her CYP2C9 status, allowed the correct warfarin dosing regimen to be achieved more quickly, thus reducing the risk of clotting or bleeding complications during the trial dose-adjustment period [18].

The cytochrome P450 enzyme family also plays a role in the variable pharmacokinetics of antiretroviral therapy (ART) for HIV patients. Higher circulating concentrations of protease inhibitors (PIs: saquinavir, indinavir, nelfinavir) have been reported in association with isozymes CYP3A5 and CYP2C19, and with non-nucleoside reverse transcriptase inhibitors (NNRTIs: efavirenz, nevirapine) with CYP2B6. In particular, serious, even life-threatening hypersensitivity reactions can occur in individuals with certain HLA types when treated with the nucleoside reverse transcriptase inhibitor (NRTI) abacavir (HLA-B*5701, HLA-DR7, and HLA-DQ3) and the NNRTI nevirapine (HLA-Cw8). Finally, some individuals with a certain mutation in the gene coding for one of the HIV co-receptors, CCR5, the so-called CCR5 delta (Δ) 32 mutation, cannot be infected by a certain phenotype (the macrophage, nonsyncytial forming [NSI] phenotype) of HIV-1, normally seen in early HIV-1 infections, and probably responsible for the majority of HIV-1 infections [19]. Hence, in both examples here, where the ingestion of exogenous, man-made chemicals (e.g., warfarin, abacavir, and nevirapine) may lead to severe illness and even death, the use of individual gene profiling may optimize the use of such drugs and reduce the risks of serious adverse effects.

1.3 TRANSLATING GENE PROFILING AND PHARMACOGENETICS TO THE BEDSIDE

Some of the problems regarding the testing, ethics, and confidentiality of individuals for gene polymorphisms have already been mentioned in this chapter. There are also different implications of gene profiling, depending on the polymorphisms found. For example, the presence of BRCA1/2 would imply that a surgical intervention (prophylactic mastectomy and/or oophorectomy) is necessary. The effectiveness of such surgical interventions to treat these conditions "in advance" is not as controversial (since these are accepted ways to treat breast and ovarian cancer) as whether and when to perform the actual genetic testing. Similarly, with FAP, the significance and options for intervention of having the genes for FAP are similarly clear-cut, i.e., the presence of the gene is clearly associated with an increased risk of cancer, and the surgical options to remove the potentially cancerous tissue are generally well-tried and tested. The main dilemma then becomes who and when to test if there is a family history of such inherited disorders, and which of the surgical options to take if they are positive for the gene(s) [6–10].

When it comes to the use of gene-profiling for enhancing patient outcomes with less clear endpoints, the question of whether gene profiling is economically worthwhile becomes more difficult to answer, both for the patient and the pharmaceutical companies that develop the drugs [20, 21]. For the companies, one way to improve upon this cost-benefit ratio is to stratify the patients, so that those most at risk of serious complications can be tested. Pharmaceutical companies have a vested interest in reducing or preventing the number of adverse effects produced by their products, as this can seriously damage their sales, as well as increase their litigation defense costs. The aim for them is to invest enough research to avoid this, but only if the rewards (increased sales and reduced numbers of adverse drug reactions) will eventually exceed their initial expenditure. One problem with targeting certain populations with a specific gene makeup for a specific drug is that these populations tend to be small and the returns may be proportionately less. Also, such a selective, targeted approach to drug development will see the gradual disappearance of the so-called blockbuster drugs whose sales have traditionally supported large pharmaceutical companies [22].

The other problem is that there are usually multiple host factors that determine the response to any drug. Testing for CYP2C9 seems to demonstrate better outcomes for the trial patients taking warfarin, but the authors of this study, themselves, admit that other factors need to be taken into consideration, since individual responses to warfarin therapy depend on a multitude of factors—not just the CYP2C9 polymorphism: "Inter-individual variability in the response to warfarin is multifactorial, involving genetic polymorphisms CYP2C9, VKORC1, and possibly other genes, host characteristics, and environmental influences" [18].

So how far do we need to go? How many tests do we need to do to ensure that all factors have been taken into consideration before deciding on the loading dose of warfarin? Some may argue that closer monitoring of the international normalized ratio (INR) for all patients starting on warfarin for the first time (i.e., treating the patient as a "black box" and monitoring the outcome only) is a better and cheaper alternative to gene profiling. For example, the cost of testing for three common cytochrome P450 polymorphisms by one company was US$250 for each of CYP2D6, CYP2C19, and CYP2C9, or a discounted US$650 for all three tests together (2004 prices) [21]. How were improved outcomes based on gene profiling assessed? This study defined the benefits as being a shorter time required to achieve stable anticoagulation, spending more time within the therapeutic range (for anticoagulation), and experiencing less minor bleeding. However, individual patients may elect to accept a greater bleeding risk, and therefore spend more time stabilizing their warfarin dose, than paying several hundred dollars for the gene testing.

Another problem in translating individualized therapies to the bedside has already been seen with the use of traditional Chinese medicine (TCM). Standards used to evaluate the efficacy, safety, and adverse effects of Western medicine with the ultimate aim of licensing are inappropriate for assessing individualized treatments, as each of us is unique [23]. However, with gene profiling to optimize clinical care, it may eventually become acceptable to have smaller numbers of patients from a specific population be involved in a clinical trial of a drug specifically designed for that population. If this select population adopted the drug after successful trials, and if there were no other alternatives on the market for a chronic illness, that pharmaceutical company would have them as customers for life [24]. Smaller trials on select populations may also avoid inadvertent adverse effects on individuals with unsuitable genotypes for that particular drug, which may also be an advantage [25].

1.4 CONCLUDING REMARKS

Despite all of these limitations, there are some attractions to gene profiling for selective drug use in certain populations, and it has been done successfully previously. Glucose-6-phosphate dehydrogenase (G6PD) deficiency has been known for many years. People with this deficiency have an enhanced fragility of their red blood cells to certain types of drugs and foods. Exposure to these drugs and food may lead to hemolytic anemia as the red cells lyse under oxidative stress, so when certain drugs are given, such as primaquine for malaria and sulfonamide antibiotics, a test for G6PD deficiency is performed. If it is positive, then these drugs are avoided. A similar strategy has been proposed for those HIV patients who may be treated with abacavir. If these patients are HLA-B*5701 positive, then abacavir is avoided to prevent a possibly fatal hypersensitivity reaction.

For some conditions, gene-profiling will not be economically viable when offered as a funded service to patients, but individual patients may choose to have the test performed directly with a private company [21]. But then the issue of confidentiality becomes paramount, i.e., the company cannot then sell the genetic information of its clients to insurance or other companies without the explicit consent of the individual tested. Designing such informed consent forms for both research and information disclosure is a daunting task [26]. Another interesting aspect of consenting for gene profiling is that, because of the inherited link to parents and siblings, it is possible that an individual's gene profile will allow some extrapolation of the results to his or her immediate relatives [25]. For example, since HD is an autosomal dominantly transmitted gene mutation, if a child is found to have the HD gene, but neither of his or her parents develops HD during their subsequent lifetimes, then it implies that either he or she was adopted, or one of the "parents" is not his or her true biological parent. Such knowledge may be a surprise (to the parents as well as the child), and may have serious consequences for subsequent family relationships.

Although there may be multiple logistical problems to bring pharmacogenetics into everyday practice [27], medical history has shown that if there is a definite benefit to a technique for improving medical care, many of these problems will eventually be solved to a point that it can be useful in everyday medical practice.

REFERENCES

1. Ebbert, E. 1988. Who will lead human genome project? *Nature* 333: 7.
2. Palca, J. 1988. James Watson to head NIH human genome project. *Nature* 335: 193.
3. Venter, J. C., Adams, M. D., Myers, E. W., Li, P. W., Mural, R. J., Sutton, G. G., et al. 2001. The sequence of the human genome. *Science*, 291, 1304.
4. Lander, E. S., Linton, L. M., Birren, B., Nusbaum, C., Zody, M. C., Baldwin, J., et al. 2001. Initial sequencing and analysis of the human genome. *Nature* 409: 860.
5. Forsyth, L., and Woof, M. 2006. The implications of the Human Tissue Act 2004 for dentistry. *Br. Dent. J.* 201: 790.
6. Botkin, J. R., Smith, K. R., Croyle, R. T., Baty, B. J., Wylie, J. E., Dutson, D., et al. 2003. Genetic testing for a BRCA1 mutation: prophylactic surgery and screening behavior in women 2 years post testing. *Am. J. Med. Genet. A.* 118: 201.

7. Foster, C., Watson, M., Eeles, R., Eccles, D., Ashley, S., Davidson, R., et al. 2007. Predictive genetic testing for BRCA1/2 in a UK clinical cohort: three-year follow-up. *Br. J. Cancer* 96: 718.

8. Lynch, H. T., Boland, C. R., Rodriguez-Bigas, M. A., Amos, C., Lynch, J. F., and Lynch, P. M. 2007. Who should be sent for genetic testing in hereditary colorectal cancer syndromes? *J. Clin. Oncol.* 25: 3534.

9. Duncan, R. E., Gillam, L., Savulescu, J., Williamson, R., Rogers, J. G., and Delatycki, M. B. 2007. "Holding your breath": interviews with young people who have undergone predictive genetic testing for Huntington disease. *Am. J. Med. Genet. A* 143: 1984.

10. Beery, T. A., and Williams, J. K. 2007. Risk reduction and health promotion behaviors following genetic testing for adult-onset disorders. *Genet. Test.* 11: 111.

11. Llewelyn, C. A., Hewitt, P. E., Knight, R. S., Amar, K., Cousens, S., Mackenzie, J., and Will, R. G. 2004. Possible transmission of variant Creutzfeldt-Jakob disease by blood transfusion. *Lancet* 363: 417.

12. Peden, A. H., Ritchie, D. L., and Ironside, J. W. 2005. Risks of transmission of variant Creutzfeldt-Jakob disease by blood transfusion. *Folia Neuropathol.* 43: 271.

13. Hewitt, P. E., Llewelyn, C. A., Mackenzie, J., and Will, R. G. 2006. Creutzfeldt-Jakob disease and blood transfusion: results of the UK Transfusion Medicine Epidemiological Review study. *Vox Sang.* 91: 221.

14. Wroe, S. J., Pal, S., Siddique, D., Hyare, H., MacFarlane, R., Joiner, S., et al. 2006. Clinical presentation and pre-mortem diagnosis of variant Creutzfeldt-Jakob disease associated with blood transfusion: a case report. *Lancet* 368: 2061.

15. Ghani, A. C., Ferguson, N. M., Donnelly, C. A., and Anderson, R. M. 2000. Predicted vCJD mortality in Great Britain. *Nature* 406: 583.

16. Duncan, R. E., Delatycki, M. B., Collins, S. J., Boyd, A., Masters, C. L., and Savulescu, J. 2005. Ethical considerations in presymptomatic testing for variant CJD. *J. Med. Ethics* 31: 625.

17. Peden, A. H., Head, M. W., Ritchie, D. L., Bell, J. E., and Ironside, J. W. 2004. Preclinical vCJD after blood transfusion in a PRNP codon 129 heterozygous patient. *Lancet* 364: 527.

18. Caraco, Y., Blotnick, S., and Muszkat, M. 2008. CYP2C9 genotype-guided warfarin prescribing enhances the efficacy and safety of anticoagulation: a prospective randomized controlled study. *Clin. Pharmacol. Ther.* Mar. 83(3): 460–470.

19. Telenti, A., and Zanger, U. M. 2008. Pharmacogenetics of anti-HIV drugs. *Annu. Rev. Pharmacol. Toxicol.* 48: 227–256.

20. Lichter, J. B., and Kurth, J. H. 1997. The impact of pharmacogenetics on the future of healthcare. *Curr. Opin. Biotechnol.* 8: 692.

21. Phillips, K. A., Veenstra, D. L., Ramsey, S. D., Van Bebber, S. L., and Sakowski, J. 2004. Genetic testing and pharmacogenomics: issues for determining the impact to healthcare delivery and costs. *Am. J. Managed Care* 10: 425.

22. Smart, A., and Martin, P. 2006. The promise of pharmacogenetics: assessing the prospects for disease and patient stratification. *Stud. Hist. Philos. Biol. Biomed. Sci.* 37: 583.

23. Jiang, W. Y. 2005. Therapeutic wisdom in traditional Chinese medicine: a perspective from modern science. *Trends Pharmacol. Sci.* 26: 558.

24. Swen, J. J., Huizinga, T. W., Gelderblom, H., de Vries, E. G. E., Assendelft, W. J. J., Kirchheiner, J., et al. 2007. Translating pharmacogenomics: challenges on the road to the clinic. *PLoS Med.* 4: e209.

25. van Delden, J., Bolt, I., Kalis, A., Derijks, J., and Leufkens, H. 2004. Tailor-made pharmacotherapy: future developments and ethical challenges in the field of pharmacogenomics. *Bioethics* 18: 303.

26. Anderson, D. C., Gomez-Mancilla, B., Spear, B. B., Barnes, D.M., Cheeseman, K., Shaw, P.M., et al. 2002. Elements of informed consent for pharmacogenetic research; perspective of the pharmacogenetics working group. *Pharmacogenomics J.* 2: 284.
27. Webster, A., Martin, P., Lewis, G., and Smart, A. 2004. Integrating pharmacogenetics into society: in search of a model. *Nat. Rev. Genet.* 5: 663.

2 Virally Encoded MicroRNA (miRNA)

Candidates for Gene Silencing

Ralph A. Tripp, Artem S. Rogovskyy,
Les P. Jones, and Stephen Mark Tompkins
University of Georgia, NIAID Center of Excellence for
Influenza Research and Surveillance, and Department
of Infectious Diseases, Animal Health Research Center

CONTENTS

2.1   INTRODUCTION

MicroRNAs (miRNAs) are small noncoding RNAs regulating gene expression through mRNA degradation or translation inhibition [1–3]. MicroRNAs were first discovered in the nematode, *Caenorhabditis elegans*, and then identified in numerous plant and animal species [2, 4–7]. A miRNA is defined as a single-stranded RNA 19–25 nucleotides (nt) in length generated from endogenous transcripts that form local hairpin structures [1, 8, 9]. The majority of miRNA genes are located in intergenic regions or in antisense orientation to annotated genes, suggesting the formation of independent transcription units [10]. Other miRNA genes are found in intronic regions that may be transcribed as part of the annotated genes [11, 12]. In animals, miRNAs are transcribed as long primary transcripts (primiRNAs) and are formed into hairpin-shaped premiRNAs by the nuclear RNase III enzyme, Drosha [13]. Drosha cleavage predetermines mature miRNA sequence and generates an optimal substrate [14]. The processing intermediate, pre-miRNA, is exported out of the nucleus by exportin-5 [15–17], where pre-miRNA is cleaved by the cytoplasmic RNase III, Dicer, into a 21–23-nt miRNA duplex [18–20]. One strand of the duplex is degraded, leaving a mature miRNA. The strand degraded is thought to be determined by the internal stability of the two ends of the duplex, i.e., the strand having the less stable 5′ end usually survives [21]. The 3′ untranslated regions (3′ UTRs) of mRNAs are targeted by miRNAs with which they share sequence complementarity [22, 23]. A summary of the steps involved in the formation of miRNAs is shown in Figure 2.1.

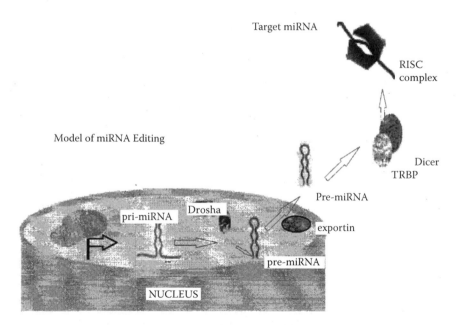

FIGURE 2.1 To generate miRNAs, a long RNA transcript, termed pri-miRNA, is initially processed into a ≈60–110-nt hairpin RNA precursor (pre-miRNA) by the nuclear ribonuclease Drosha. The resultant pre-miRNA is processed by Dicer to generate a ≈22-bp duplex. Subsequently, one strand of the duplex is incorporated into the RNA-induced silencing complex (RISC) of the RNAi pathway and guides the RISC to the target transcript that is complementary to the miRNA sequence. RISC binding results in either mRNA cleavage or translation inhibition, depending on the degree of complementarity between the miRNA and the target transcript.

2.2 ROLE OF miRNAS

Recent studies have revealed that miRNAs have key roles in regulatory pathways, including development, apoptosis, cell proliferation and differentiation, organ development, and cancer [2, 24–27]. Release 9.1 of miRBase contains 1993 mammalian miRNA sequences, including 377 murine and 474 human sequences (May 2007) [28, 29]. It has been estimated that there are approximately 200 predicted conserved mRNA targets comprising more than 5300 human genes per mammalian miRNA, suggesting that up to 30% of the human genome may be regulated by miRNAs [30]. These findings suggest that miRNAs may act to fine-tune gene expression and function as complex regulators of cellular networks. Importantly, it has been established that a single miRNA can bind to and regulate many different mRNA targets, and that several different miRNAs can bind to and cooperatively control a single mRNA target [22, 31]. The recently recognized role of miRNAs in regulating aspects of hematopoietic cell proliferation and differentiation has led researchers to investigate the role of miRNAs in cancer predisposition. It appears that some miRNAs may serve as biomarkers for disease, particularly in the development of some cancers. Expression studies of various tumor types have revealed specific alterations in miRNA profiles. For example, mir-15 and mir-16 are frequently deleted and/or down-regulated in B-cell chronic lymphocytic leukemia [32, 33], and levels of miR-143 and miR-145 are reduced in colorectal cancer [34]. Recent studies suggest that clusters of miRNAs may also act as potent oncogenes. For example, miRNA clusters on human chromosomes (MIRN17-92) have been found to be regulated by transcription factors, e.g., c-Myc, that are overexpressed in many human cancers [35–38].

2.2.1 MIRNA REGULATION

MicroRNAs are not uniformly expressed in host cells. Mammalian miRNAs can be expressed in a tissue-specific and/or developmental-stage-specific manner. Like mRNA, many miRNAs are regulated by a variety of mechanisms, including miRNA-specific promoter-regulated expression, regulation of host genes containing miRNAs in their exons and introns, regulation of expression of noncoding mRNAs containing miRNAs, and posttranscriptional regulation [39, 40]. While most mammalian miRNAs are expressed as individuals, many are expressed in clusters under the control of the same promoter [41]. Lastly, many miRNAs are located at fragile sites in the genome and are implicated in oncogenesis [42, 43].

While *C. elegans* and *Drosophila melanogaster* are classical developmental models, murine and other rodent models have been utilized to characterize miRNAs' function in mammals. Initial studies deleting miRNA processing machinery demonstrated a critical role for miRNAs in development. Deletion of *Dicer1* resulted in lethality early during embryonic development and an inability to generate stem cells, suggesting that miRNAs are critical to stem cell development [44]. Utilizing conditional mutation systems, *Dicer1* was shown to be critical for stem cell proliferation and differentiation and for centromere function [45, 46]. Similarly, loss of Argonaute-2 (Ago2) produced a late-embryonic-lethal phenotype, which showed cardiac and neural tube defects [47]. Taken together, the requirement for miRNA function demonstrates a crucial role for miRNA activity in mammalian development.

Tissue-specific regulation of miRNA expression and function, and expression profiling of miRNAs, has more clearly delineated the many diverse roles for miRNAs.

2.2.2   miRNA and Host Genes

A variety of miRNAs that regulate late stages of development have been identified. miR-196 is expressed during mouse embryogenesis, regulating the expression of HOXB8 mRNA that encodes a transcription factor crucial for a variety of developmental programs in animals [48, 49]. Ectopic expression of miR-181 in hematopoietic stem cells was shown to increase the fraction of B-lineage cells, and ectopic expression of miR-142 or miR-223 increased the fraction of T-lineage cells [50]. miR-1-1 and miR-1-2 were shown to be expressed in cardiac and skeletal muscle precursor cells and to regulate expression of the cardiac transcription factor, Hand2 [51]. Here, the miR-1 regulation appeared to act both spatially and temporally to fine-tune cardiogenesis. More recently, miR-141, miR-199a, miR-200a, and miR-429 were shown to regulate skin morphogenesis [52]. A variety of miRNAs have been identified in the central nervous system and have been implicated in neuronal development. miR132 was enriched in neurons and was overexpression induced, while inhibition attenuated neuronal outgrowth. miR132 expression was regulated by a cAMP-response element binding protein (CREB) and in turn modulated neuronal morphogenesis by decreasing levels of p250GAP, a GTPase-activating protein [53]. Another miRNA, miR-134, was found specifically in rat brain tissue, expressed locally in the dendrites of neurons, and shown to negatively regulate dendritic development [54].

In addition to developmental regulation, miRNAs have roles in a variety of cellular processes. miR-122 has been identified as a liver-specific miRNA [55], and data suggest that it may specifically target and regulate CAT-1, a cationic amino acid transporter, and result in the lack of CAT-1 protein exclusively in the liver [56]. Using antisense oligonucleotides to reduce miRNA expression in hepatocytes, miR-122 was also shown to regulate cholesterol and fatty-acid metabolism [57]. miR-375 was shown to be expressed in murine pancreatic islets, and its overexpression suppressed glucose-induced insulin secretion, while inhibition of miR-375 enhanced insulin secretion [58].

2.2.3   miRNAs and Oncogenesis

The multifaceted role that miRNAs have in cellular development and physiological processes has led to the investigation of aberrant miRNA expression influencing oncogenesis. There is now clear evidence that miRNAs can act as tumor suppressors and oncogenes. For example, human miRNAs miR-15 and miR-16 are deleted or down-regulated in most B-cell chronic lymphocytic leukemia [59], suggesting that these miRNAs are acting as tumor-suppressors. Similarly, let-7 miRNA is down-regulated in lung cancer [60]. In contrast, the hairpin precursor of miR-155 was found in abundance in 100% of pediatric Burkett lymphoma patients, but not in other leukemia [61], suggesting miR-155 has oncogenic potential. Likewise, miRNA miR-21 has been shown to be overexpressed in human glioblastoma tumor tissues, in early-passage glioblastoma cultures, and in established glioblastoma cell lines [62]. Importantly, inhibition of miR-21 has been linked to an increase in apoptotic cell death in cultured glioblastoma cells [62], suggesting a potential therapeutic approach. The findings that miRNAs are associated with cancer suggest that miRNA expression profiles may be useful to classify and diagnose human cancers. In fact, cancer-specific miRNA expression patterns have been identified in every cancer analyzed to date (reviewed by Calin and Croce [63]).

2.2.4 MIRNAS AND THE IMMUNE RESPONSE

In addition to influencing B- and T-cell hematopoiesis, miRNAs have been shown to have many roles in the development and regulation of the immune response. Using microarray expression analysis, 181 miRNAs were screened for expression during discrete stages of hematopoiesis, ranging from progenitor cells to differentiated B cells, T cells, and macrophages [64]. miR-150 was found to be up-regulated during the early stages of B- and T-cell maturation, but was down-regulated in differentiated T-helper (Th1) and Th2 cells. In contrast, miR-146 was found to be up-regulated only after Th1 cell differentiation. Studies examining the role of a conditional mutation of *Dicer1* provided evidence that miRNA expression was important for the generation and survival of α/β T cells, but not for CD4$^+$ or CD8$^+$ T-cell commitment [65]. miRNAs have also been shown to have a critical role in immune regulation. CD4$^+$ regulatory T (T-reg) cells were found to have distinct miRNA expression profiles compared with CD4$^+$ T helper (Th) cells [66], supporting the hypothesis that T-reg cells are maintained at a level of partial activation. miRNAs have also been shown to have a role in regulating signal transduction during immune responses. miR-146, miR-132, and miR-155 were shown to be up-regulated in human monocytes following lipopolysaccharide stimulation [67]. In these studies, miR-146 expression was found to be NF-κB-dependent and induced by Toll-like receptor (TLR) signaling. Recently, miR-155 was found to be up-regulated in murine bone-marrow-derived macrophages after poly-(I:C) or IFNγ stimulation [68].

2.3 VIRUS REGULATION

In addition to regulating lymphocyte development and signaling, miRNAs have been shown to have direct antiviral activity. For example, miR-132 was shown to limit the replication of primate foamy virus type 1 (PFV-1) [69]. In these studies, miR-132 was shown to target regions in open reading frame (ORF) 2 and within 3'UTRs of PVF-1. Treatment of cells with miR-132 inhibitors, or infection with mutant viruses lacking the miR-132 target sites, increased PVF-1 viral load two- to threefold [69]. In contrast, miR-122 was shown to modulate hepatitis C virus RNA abundance, its absence reducing HCV RNA levels in infected cells [70]. Using computational analysis, five human miRNAs (miR-29a/b, miR-149, miR-324-5p, and miR-378) were proposed to have a number of potential targets in the HIV genome [71], and, using similar methods, potential binding sites for human miRNAs miR-507 and miR-136 were identified in the basic polymerase 2 (PB2) and hemagglutinin (HA) genes of influenza A virus, respectively [72]. Thus, miRNAs may be directly acting on viral pathogens during infection.

2.3.1 VIRALLY ENCODED miRNAs

Recent evidence suggests that virally encoded miRNAs (or siRNAs [small interfering RNA] that may arise through the miRNA biogenesis pathway) may have important roles for virus gene expression and the regulation of the antiviral or host cell response to infection [73, 74]. Virally encoded miRNAs may be important to facilitate replication, as the host cell has numerous antiviral defense mechanisms, e.g., RNA interference, apoptosis, type I IFNs, Toll-like receptors, etc. [75–80], and it is likely that host cell miRNAs impinge on aspects of the viral life cycle, cell tropism, and the pathogenesis of viral diseases as well. Further, we hypothesize here that virally encoded miRNAs may be critical in regulating both viral and host gene expression in a temporal and tissue-specific manner in order to facilitate replication. Thus we would predict that virally encoded miRNAs may be key to the systemic spread of virus infection and development of virus persistence or latency.

The first examples of virally encoded miRNAs were reported in B-cell lines latently infected with Epstein–Barr virus (EBV) [81]. However, potential miRNAs have been linked to several other latent viruses, including HIV-1 [82], Kaposi sarcoma-associated virus (KSHV or HHV8), mouse gammaherpesvirus 68 (MHV68), and human cytomegalovirus (HCMV or HHV5) [81]. It is important to note that, according to the known miRNA pathway, not all viruses should have the potential to generate miRNAs because the initial step in miRNA processing, i.e., cleavage of the pri-miRNA by Drosha, occurs in the nucleus [13, 83, 84]. This is particularly true for most RNA viruses that replicate and express their genome in the cytoplasm [85, 86]. However, pre-miRNA generation relies on export to the cytoplasm, where Dicer and TRBP interact to produce miRNA for assembly with the multiprotein RNA induced silencing complex (RISC). Therefore, it is possible that during infection virally encoded miRNAs may accumulate in infected cells, where they may interact with Dicer and TRBP, or be made immediately available to associate with RISC. Nonetheless, it may not be necessary for viruses using lytic replication methods to encode miRNAs given their relatively short life cycle. For example, miRNAs act exclusively on mRNAs [87–89] and cannot exert a marked phenotypic effect until a preexisting host cell encoded protein has decayed. Thus, miRNAs may not be useful to augment lytic replication for some viruses, as they may not require the time needed to modify the host cell environment to facilitate replication.

It is important to note that some RNA viruses that establish persistent infection, such as retroviruses, e.g., HIV-1, have the ability to integrate into the host genome as a provirus [90–92]. Since the provirus is used as a template for gene expression, and all transcription originates from host cell machinery, it is likely that this machinery is also used to direct expression of viral miRNAs. Several predicted HIV pre-miRNA stem-loop structures have been proposed to function as siRNAs [93, 94], and a novel miRNA, miR-N367, has been implicated in the Nef region of the genome [95–97]. It has been shown that miR-N367 can block HIV-1 Nef expression and long-terminal repeat transcription, suggesting that miR-N367 might suppress HIV-1 transcription through the RNAi pathway [95]. Thus, for viruses that establish latent infections, virally encoded miRNAs may help to diminish a hostile host cell environment, making it more suitable for virus replication.

2.4 miRNA CANDIDATES IN RESPIRATORY SYNCYTIAL VIRUS (RSV) AND INFLUENZA VIRUS

Viruses are adept at capturing host genes and using their genes to modify host cell responses [98–102]. Given the growing body of evidence implicating miRNAs as key regulators of a spectrum of host cell pathways and gene expression, and the indications that miRNAs are highly conserved [103–105], it is little wonder that viruses may have usurped miRNAs as a method to facilitate infection likely through regulation of their own genes and those of the host cell. There is a growing list of putative and some known virally encoded miRNAs [106]. Many of these miRNAs have been shown to target the expression of a wide range of antiviral host cell genes, among them genes for cytokines and signaling proteins [73, 107]. The majority of virally encoded miRNAs have been identified as homologues to validated miRNAs using alignment programs. However, sequence alignment may not be sufficient to identify virally encoded miRNAs that have diverged in primary sequence but still retain their Watson-Crick base-pairing ability. To overcome aspects of this limitation, computational techniques can be exploited to analyze both RNA structure and sequence alignments for noncoding RNA searches [22, 108–110]. These techniques provide a probabilistic description of RNA sequences and structure. However, the Easy RNA Profile IdentificatioN (ERPIN) program can be used as an RNA motif identification program that takes an RNA sequence alignment as an input and identifies related sequences using a profile-based dynamic programming algorithm [111]. ERPIN differs from other RNA motif search programs because it captures both primary and secondary structure information. This feature makes ERPIN useful for miRNA precursor identification [111]. ERPIN profiles miRNA motif identification by providing E-value calculations, which are calculated based on a discrete convolution analysis of profile scores. ERPIN-detected miRNA candidates with lower E-values suggest the presence of strong miRNA candidates that are likely unannotated in current databases.

2.4.1 ALGORITHM PREDICTION OF miRNAS

We chose to identify potential miRNA candidates in the genomes of two important human respiratory viruses, i.e., respiratory syncytial virus (RSV) and influenza virus (influenza). Influenza is a negative-strand, RNA virus in the *Orthomyxoviridae* family that may cause severe respiratory disease in animals and humans [112]. Influenza is classified into three families, types A, B, and C. Types A and C may infect a variety of species, including humans and birds, while type B viruses infect only humans [113, 114]. The influenza genome consists of single-strand negative-sense RNA in eight segments (seven in influenza C). Each RNA segment encodes one or two proteins, and the gene assignment is as follows: polybasic protein 2 (PB2), PB1, acidic polymerase (PA), hemagglutinin (HA), nucleoprotein (NP), neuraminidase (NA), matrix protein-1 (M1) and M2, and nonstructural protein-1 (NS1) and NS2 [113, 114]. RSV is a single-stranded, negative-sense RNA virus in the Paramyxoviridae family and is the single most important cause of serious lower respiratory tract disease in infants and young children worldwide [102]. There are two major groups of RSV, strains A and B. The RSV genome contains two nonstructural (NS1/NS2) genes followed by nucleocapsid (N), phosphoprotein (P), matrix (M), small hydrophobic (SH), attachment (G), fusion (F), second matrix (M2), and RNA-dependent RNA polymerase (L) genes. The gene order is 3'-NS1-NS2-N-P-M-SH-G-F-M2-L-5' [102]. We used a procedure based on CLUSTALW alignment [115] and ERPIN analyses [111] to scan the entire genomes of RSV strain A2 and influenza A/Puerto Rico/8/34/Mount Sinai (H1N1). The analyses detected numerous hits for each virus examined; however, 14 hits with relatively low E-values (≤4) were detected for influenza and 11 hits detected for RSV (see Tables 2.1 and 2.2, respectively). The hits were then compared with known annotated miRNAs and the predicted virally encoded miRNA targets determined.

2.4.2 INFLUENZA miRNA CANDIDATES

The miRNA matches based on relatively low E-values for influenza virus (Table 2.1) reveal a diverse number of conserved host gene targets for the putative virally encoded miRNAs. Interestingly, these supposed virally encoded miRNAs span the influenza genome: one in NA and NS1/NS2; two in PB2, HA, and NP; and three in PB1 and PA. The number of conserved host gene targets for each viral miRNA varies from 3 (MiR-471) to 651 (MiR-424). Of the potential host genes targeted, several putative miRNAs have predicted host gene targets that seem critical for influenza replication. For example, miR-351, miR-424, and let-7c are predicted to target IFN regulatory factors, IL-10, transcription factors, ion channels, suppressor of cytokine signaling (SOCS) proteins, cell cyclins, and apoptosis, whereas miR150, miR-150, miR-202, and miR-541 are predicted to target membrane glycoproteins, mannose receptors, solute carrier family members, and translation initiation factors.

2.4.3 RSV miRNA CANDIDATES

RSV also encodes miRNAs spanning the RSV genome, but there is a trend toward miRNAs locating in nonessential genes (Table 2.2). For example, four putative miRNAs are located in NS1/NS2; three in G; and one in NP, P, M, and 22K/M2. It has been shown that both the NS1/NS2 genes and RSV G gene are not required for RSV replication [102, 116–119]. However, these genes are conserved in all RSV strains and isolates, suggesting that these genes may have more than an auxiliary function for virus replication that may be linked in part to miRNA activity. As for influenza, the RSV-encoded miRNAs have a number of conserved host gene targets. However, most of the putative miRNAs are predicted to target cytokine and chemokine expression as well as immune activation. For example, miR-142, miR-197, miR-291, miR-146, and miR-369 may target Toll/IL-1 receptors, IL-6 signal transduction, TNF receptor, IL-10, and IL-15 receptors, as well as monocyte-to-macrophage differentiation, B-cell translocation, CXC and CX2C chemokines, and immunoglobulin superfamily members. Some of the other miRNAs, such as miR-142, miR-503, and miR-337, target transcription activators, nuclear factors, cell division, and transmembrane proteins.

TABLE 2.1
miRNA Matches for Influenza A Virus (A/Puerto Rico/8/34/Mount Sinai)

Putative miRNA	Influenza Viral Gene	Number of Conserved Targets	Predicted miRNA Targets
miR-542-5p	segment 1 (PB2)	5	Transcription factor, integrin, protein peptidase
miR-150	segment 1 (PB2)	165	Membrane glycoproteins, nuclear receptor subfamily members, transcription factors, mannose receptors, eukaryotic translation initiation factors, solute carrier family members
miR-351	segment 2 (PB1)	165	IFN regulatory factor, regulatory factors, zinc finger proteins, transcription factors, hormone receptors, mannose receptor, nucleoporin, suppressor of cytokine signaling, cyclins
miR-490	segment 2 (PB1)	104	Translation termination factors, zinc finger proteins, solute carrier family members, ion channels, cyclin, protein phosphatase regulatory subunit
miR-541	segment 2 (PB1)	63	Zinc finger proteins, cytoplasmic polyadenylation element binding protein, thiamine triphosphatase, solute carrier family members, guanine nucleotide binding protein, translation initiation factors, transmembrane proteins, integrin-linked kinase
miR-202	segment 3 (PA)	427	Cyclin, interleukin-10, zinc finger members, cell receptors, transcription factors, solute carrier family transmembrane proteases, oncogenes
miR-1	segment 3 (PA)	440	Macrophage differentiation, neurotropic factor, zinc finger proteins, nuclear factors, ion channel, transcription factor, transmembrane proteins, protein inhibitor, retinoic acid receptor
miR-511	segment 3 (PA)	267	Solute carrier family, cadherin, cyclins, calmodulin-dependent protein kinase II inhibitor, translation initiation factor, suppressor of cytokine signaling, transcription factor, transmembrane proteins, ion channel members, splicing factor
miR-518c	segment 4 (HA)	25	Ion channel member, nuclear factor, oncogene family member, enhancer of transcription factor, membrane-bound transcription factor peptidase
miR-136	segment 4 (HA)	125	Transcription complex subunit, translation termination factor, growth differentiation factor, phospholipases, proteasome activator subunit, ion channel member, splicing factors, apoptosis inhibitor
let-7c	segment 5 (NP)	561	Interleukin 10, interleukin 6, cyclins, transcriptional factors, zinc finger proteins, ion channel domains, immunoglobulin-like domains, suppressors of cytokine signaling, TNF receptor superfamily

TABLE 2.1 (CONTINUED)
miRNA Matches for Influenza A Virus (A/Puerto Rico/8/34/Mount Sinai)

Putative miRNA	Influenza Viral Gene	Number of Conserved Targets	Predicted miRNA Targets
miR-424	segment 5 (NP)	651	Cyclins, suppressor of cytokine signaling, growth factors, enzymes, ion channels, cell receptors, transcription factors, growth hormone receptor, RNA binding motif protein, calcium binding protein, translation initiation factors, transmembrane proteins
miR-471	segment 6 (NA)	3	Chromatin regulator
miR-518c	segment 8 (NS1, NS2)	25	Zinc finger proteins, nuclear factor, tumor proteins, transcription factors, enzymes

TABLE 2.2
miRNA Matches for Respiratory Syncytial Virus (RSV) Strain A2

Putative miRNA	RSV Viral Gene	Number of Conserved Targets	Predicted miRNA Targets
miR-142-5p	NS2/1B	422	Transcription activators, nuclear factor, cyclins, ion channels, chemokine orphan receptor, transcription factor
miR-142-3p	NS2/1B	172	Interleukin-6 signal transducer, interleukin-1 receptor-associated kinase, guanine nucleotide binding proteins, retinoic acid receptor, transcription factors, transmembrane proteins, clathrin, growth factors, monocyte-to-macrophage differentiation-associated genes, transmembrane proteins, interleukin 1 receptor accessory protein, nuclear receptor subfamily members
miR-291a-3p	NS2/1B	391	Tumor necrosis factor, interferon regulatory factor, immunoglobulin superfamily members, transcription complex subunit, transcription factors, nuclear receptor subfamily members, nuclear receptor coactivator, cyclins, ubiquitin-conjugating enzymes, ion channels
miR-291-3p	NS2/1B	391	B-cell translocation gene, transcription factor, nuclear receptor subfamily members, tumor necrosis factor, ubiquitin-conjugating enzymes, cyclins, mitochondrial ribosomal protein, chemokine (C-X-C motif) receptor, ion channel
miR-503	NP	259	Cyclins, cell-division-cycle associated, ion channels, solute carrier family members, transmembrane protein, ubiquitination factor, zinc finger proteins

TABLE 2.2 (CONTINUED)
miRNA Matches for Respiratory Syncytial Virus (RSV) Strain A2

Putative miRNA	RSV Viral Gene	Number of Conserved Targets	Predicted miRNA Targets
miR-146	P	101	Immunoglobulin superfamily members, interleukin-1 receptor-associated kinase, retinoic acid receptor, TNF receptor-associated factor, immunoglobulin-like variable motif, proteasome, tubulin, early growth response, eukaryotic translation initiation factor
miR-291a-3p	M	391	Transcription complex subunit, nuclear receptor subfamily members, tumor necrosis factor, B-cell translocation gene, interferon regulatory factor 2, cyclins, cell division cycle associated, CXXC, immunoglobulin superfamily members
miR-369-3p	G	377	Immunoglobulin superfamily members, interleukin 10, interleukin 15 receptor, protein kinases, zinc fingers, transcription factors, calcium-sensing receptor, cell division cycle, ion channels, splicing factors, eukaryotic translation initiation factors, ion channels, solute carrier family members, transmembrane proteins
miR-337	G	224	Retinoic acid receptor, splicing factors, nuclear receptor subfamily members, solute carrier family, ubiquitin-specific peptidase, ubiquitin-conjugating enzyme
miR-197	G	224	Splicing factors, retinoic acid receptor, ion interferon regulatory factor, channels, cytokine-like nuclear factor, platelet-derived growth factor receptor, tumor necrosis factor receptor superfamily
miR-186	22K/L gene overlap region	447	Transcription factors, immunoglobulin superfamily members, integrins, translocation protein, insulin-like growth factor, microtubule-associated protein, neuronal growth regulator, calcium channel, eukaryotic translation initiation factors, nuclear receptor coactivator, proteasome subunits

2.5 CONCLUDING REMARKS

The predicted host gene targets have commonalities for influenza and RSV miRNAs and suggest potential roles for viral or host gene regulatory pathways. It is possible that virally encoded miRNAs targeting transcription or nuclear factors may be more important for viral gene regulation than host gene regulation, as RSV and influenza utilize lytic replication strategies. In contrast, those viral miRNAs targeting transmembrane proteins, ion channels, SOCS, cytokines, chemokines, or cell surface receptors target host genes to facilitate virus replication. It is clear that validation of the role of putative viral miRNAs is required, as understanding their biological role may modify the dogma and our understanding of virus replication, host gene regulation, and the development of live-attenuated vaccines as well as disease intervention strategies.

Various methods have been used to approach the validation of virally encoded miRNAs [23, 120–122], a process that generally involves identifying the miRNA sequences and determining their expression. Alternative approaches involve the use of "antagomirs," which are a class of chemically engineered oligonucleotides that specifically and efficiently silence endogenous miRNAs [123–126]. As Watson-Crick base-pair complementarity is required between miRNAs and mRNAs for the function of miRNAs [21, 30], antagomirs are effective because they act to inhibit such interactions by complementary binding of an antisense oligonucleotide to the miRNA. Thus, specific removal of the target miRNAs by antagomirs can be used to confirm the gain or loss of phenotype associated with putative miRNAs. Taken together, there are sufficient tools for defining and validating virally encoded miR-NAs, as well as host miRNAs, and these can be used to advance the infant field of miRNA biology.

REFERENCES

1. Scherr, M., and Eder, M. 2004. RNAi in functional genomics. *Curr. Opin. Mol. Ther.* 6: 129.
2. Bartel, D. P. 2004. MicroRNAs: genomics, biogenesis, mechanism, and function. *Cell* 116: 281.
3. Zhang, B., Wang, Q., and Pan, X. 2007. MicroRNAs and their regulatory roles in animals and plants. *J. Cell. Physiol.* 210: 279.
4. Ying, S. Y., and Lin, S. L. 2005. Intronic microRNAs. *Biochem. Biophys. Res. Commun.* 326: 515.
5. Pfeffer, S., Zavolan, M., Grasser, F. A., Chien, M., Russo, J. J., Ju, J., John, B., Enright, A. J., Marks, D., Sander, C., and Tuschl, T. 2004. Identification of virus-encoded microRNAs. *Science* 304: 734.
6. Watson, J. M., Fusaro, A. F., Wang, M., and Waterhouse, P. M. 2005. RNA silencing platforms in plants. *FEBS Lett.* 579: 5982.
7. Wang, M. B., and Metzlaff, M. 2005. RNA silencing and antiviral defense in plants. *Curr. Opin. Plant Biol.* 8: 216.
8. Chang, K., Elledge, S. J., and Hannon, G. J. 2006. Lessons from nature: microRNA-based shRNA libraries. *Nat. Methods* 3: 707.
9. Murchison, E. P., and Hannon, G. J. 2004. miRNAs on the move: miRNA biogenesis and the RNAi machinery. *Curr. Opin. Cell. Biol.* 16: 223.
10. Lau, N. C., Lim, L. P., Weinstein, E. G., and Bartel, D. P. 2001. An abundant class of tiny RNAs with probable regulatory roles in *Caenorhabditis elegans. Science* 294: 858.
11. Ambros, V., and Lee, R. C. 2004. Identification of microRNAs and other tiny noncoding RNAs by cDNA cloning. *Methods Mol. Biol.* 265: 131.
12. Rodriguez, A., Griffiths-Jones, S., Ashurst, J. L., and Bradley, A. 2004. Identification of mammalian microRNA host genes and transcription units. *Genome Res.* 14: 1902.
13. Lee, C. T., Risom, T., and Strauss, W. M. 2006. MicroRNAs in mammalian development. *Birth Defects Res. C Embryo Today.* 78: 129.
14. Lee, Y., Ahn, C., Han, J., Choi, H., Kim, J., Yim, J., Lee, J., Provost, P., Radmark, O., Kim, S., and Kim, V. N. 2003. The nuclear RNase III Drosha initiates microRNA processing. *Nature* 425: 415.
15. Yi, R., Qin, Y., Macara, I. G., and Cullen, B. R. 2003. Exportin-5 mediates the nuclear export of pre-microRNAs and short hairpin RNAs. *Genes Dev.* 17: 3011.

16. Lund, E., and Dahlberg, J. E. 2006. Substrate selectivity of exportin 5 and Dicer in the biogenesis of microRNAs, *Cold Spring Harb. Symp. Quant. Biol.* 71: 59.

17. Lund, E., Guttinger, S., Calado, A., Dahlberg, J. E., and Kutay, U. 2004. Nuclear export of microRNA precursors. *Science* 303: 95.

18. Bernstein, E., Caudy, A. A., Hammond, S. M., and Hannon, G. J. 2001. Role for a bidentate ribonuclease in the initiation step of RNA interference. *Nature* 409: 363.

19. Bernstein, E., Kim, S. Y., Carmell, M. A., Murchison, E. P., Alcorn, H., Li, M. Z., Mills, A. A., Elledge, S. J., Anderson, K. V., and Hannon, G. J. 2003. Dicer is essential for mouse development. *Nat. Genet.* 35: 215.

20. Denli, A. M., Tops, B. B., Plasterk, R. H., Ketting, R. F., and Hannon, G. J. 2004. Processing of primary microRNAs by the microprocessor complex. *Nature* 432: 231.

21. Schwarz, D. S., Hutvagner, G., Du, T., Xu, Z., Aronin, N., and Zamore, P. D. 2003. Asymmetry in the assembly of the RNAi enzyme complex. *Cell* 115: 199.

22. Lewis, B. P., Shih, I. H., Jones-Rhoades, M. W., Bartel, D. P., and Burge, C. B. 2003. Prediction of mammalian microRNA targets. *Cell* 115: 787.

23. Hsu, P. W., Lin, L. Z., Hsu, S. D., Hsu, J. B., and Huang, H. D. 2007. ViTa: prediction of host microRNAs targets on viruses. *Nucleic Acids Res.* 35 (database issue): D381.

24. Bartel, B. 2005. MicroRNAs directing siRNA biogenesis. *Nat. Struct. Mol. Biol.* 12: 569.

25. Yeung, M. L., Bennasser, Y., and Jeang, K. T. 2007. miRNAs in the biology of cancers and viral infections. *Curr. Med. Chem.* 14: 191.

26. Pfeffer, S., and Voinnet, O. 2006. Viruses, microRNAs and cancer. *Oncogene* 25: 6211.

27. Nair, V., and Zavolan, M. 2006. Virus-encoded microRNAs: novel regulators of gene expression. *Trends Microbiol.* 14: 169.

28. Griffiths-Jones, S., Grocock, R. J., van Dongen, S., Bateman, A., and Enright, A. J. 2006. miRBase: microRNA sequences, targets and gene nomenclature. *Nucl. Acids Res.* 34 (Suppl. 1): D140.

29. Griffiths-Jones, S. 2004. The microRNA registry. *Nucl. Acids Res.* 32 (Suppl. 1): D109.

30. Lewis, B. P., Burge, C. B., and Bartel, D. P. 2005. Conserved seed pairing, often flanked by adenosines, indicates that thousands of human genes are microRNA targets. *Cell* 120: 15.

31. Farh, K. K., Grimson, A., Jan, C., Lewis, B. P., Johnston, W. K., Lim, L. P., Burge, C. B., and Bartel, D. P. 2005. The widespread impact of mammalian MicroRNAs on mRNA repression and evolution. *Science* 310: 1817.

32. Calin, G. A., and Croce, C. M. 2006. Genomics of chronic lymphocytic leukemia microRNAs as new players with clinical significance. *Semin. Oncol.* 33: 167.

33. Cimmino, A., Calin, G. A., Fabbri, M., Iorio, M. V., Ferracin, M., Shimizu, M., Wojcik, S. E., Aqeilan, R. I., Zupo, S., Dono, M., Rassenti, L., Alder, H., Volinia, S., Liu, C. G., Kipps, T. J., Negrini, M., and Croce, C. M. 2005. miR-15 and miR-16 induce apoptosis by targeting BCL2, *Proc. Natl. Acad. Sci. USA* 102: 13944.

34. Michael, M. Z., O'Connor, S. M., van Holst Pellekaan, N. G., Young, G. P., and James, R. J. 2003. Reduced accumulation of specific microRNAs in colorectal neoplasia. *Mol. Cancer Res.* 1: 882.

35. Gaur, A., Jewell, D. A., Liang, Y., Ridzon, D., Moore, J. H., Chen, C., Ambros, V. R., and Israel, M. A. 2007. Characterization of microRNA expression levels and their biological correlates in human cancer cell lines. *Cancer Res.* 67: 2456.

36. Hammond, S. M. 2006. RNAi, microRNAs, and human disease. *Cancer Chemother. Pharmacol.* 58 (Suppl. 1): s63.

37. He, L., Thomson, J. M., Hemann, M. T., Hernando-Monge, E., Mu, D., Goodson, S., Powers, S., Cordon-Cardo, C., Lowe, S. W., Hannon, G. J., and Hammond, S. M. 2005. A microRNA polycistron as a potential human oncogene. *Nature* 435: 828.

38. Tagawa, H., and Seto, M. 2005. A microRNA cluster as a target of genomic amplification in malignant lymphoma. *Leukemia* 19: 2013.
39. Di Leva, G., Calin, G. A., and Croce, C. M. 2006. MicroRNAs: fundamental facts and involvement in human diseases. *Birth Defects Research Part C: Embryo Today: Reviews* 78: 180.
40. Obernosterer, G., Leuschner, P. J. F., Alenius, M., and Martinez, J. 2006. Post-transcriptional regulation of microRNA expression. *RNA* 12: 1161.
41. Conrad R., Barrier, M., and Ford, L. P. 2006. Role of miRNA and miRNA processing factors in development and disease. *Birth Defects Res. Part C: Embryo Today* 78: 107.
42. Calin, G. A., Sevignani, C., Dumitru, C. D., Hyslop, T., Noch, E., Yendamuri, S., Shimizu, M., Rattan, S., Bullrich, F., Negrini, M., and Croce, C. M. 2004. Human microRNA genes are frequently located at fragile sites and genomic regions involved in cancers. *Proc. Natl. Acad. Sci. USA* 101: 2999.
43. Huppi, K., Volfovsky, N., Mackiewicz, M., Runfola, T., Jones, T. L., Martin, S. E., Stephens, R., and Caplen, N. J. 2007. MicroRNAs and genomic instability. *Seminars in Cancer Biology* 17: 65.
44. Bernstein, E., Kim, S. Y., Carmell, M. A., Murchison, E. P., Alcorn, H., Li, M. Z., Mills, A. A., Elledge, S. J., Anderson, K. V., and Hannon, G. J. 2003. Dicer is essential for mouse development. *Nat. Genet.* 35: 215.
45. Kanellopoulou, C., Muljo, S. A., Kung, A. L., Ganesan, S., Drapkin, R., Jenuwein, T., Livingston, D. M., and Rajewsky, K. 2005. Dicer-deficient mouse embryonic stem cells are defective in differentiation and centromeric silencing. *Genes Dev.* 19: 489.
46. Murchison, E. P., Partridge, J. F., Tam, O. H., Cheloufi, S., and Hannon, G. J. 2005. Characterization of Dicer-deficient murine embryonic stem cells. *Proc. Natl. Acad. Sci. USA* 102: 12135.
47. Liu, J., Carmell, M. A., Rivas, F. V., Marsden, C. G., Thomson, J. M., Song, J.-J., Hammond, S. M., Joshua-Tor, L., and Hannon, G. J. 2004. Argonaute2 is the catalytic engine of mammalian RNAi. *Science* 305: 1437.
48. Yekta, S., Shih, I. H., and Bartel, D. P. 2004. MicroRNA-directed cleavage of HOXB8 mRNA. *Science* 304: 594.
49. Hornstein, E., Mansfield, J. H., Yekta, S., Hu, J. K.-H., Harfe, B. D., McManus, M. T., Baskerville, S., Bartel, D. P., and Tabin, C. J. 2005. The microRNA miR-196 acts upstream of Hoxb8 and Shh in limb development. *Nature* 438: 671.
50. Chen, C.-Z., Li, L., Lodish, H. F., and Bartel, D. P. 2004. MicroRNAs modulate hematopoietic lineage differentiation. *Science* 303: 83.
51. Zhao, Y., Samal, E., and Srivastava, D. 2005. Serum response factor regulates a muscle-specific microRNA that targets Hand2 during cardiogenesis. *Nature* 436: 214.
52. Yi, R., O'Carroll, D., Pasolli, H. A., Zhang, Z., Dietrich, F. S., Tarakhovsky, A., and Fuchs, E. 2006. Morphogenesis in skin is governed by discrete sets of differentially expressed microRNAs. *Nat. Genet.* 38: 356.
53. Vo, N., Klein, M. E., Varlamova, O., Keller, D. M., Yamamoto, T., Goodman, R. H., and Impey, S. 2005. From the cover: a cAMP-response element binding protein-induced microRNA regulates neuronal morphogenesis. *Proc. Natl. Acad. Sci. USA* 102: 16426.
54. Schratt, G. M., Tuebing, F., Nigh, E. A., Kane, C. G., Sabatini, M. E., Kiebler, M., and Greenberg, M. E. 2006. A brain-specific microRNA regulates dendritic spine development. *Nature* 439: 283.
55. Lagos-Quintana, M., Rauhut, R., Yalcin, A., Meyer, J., Lendeckel, W., and Tuschl, T. 2002. Identification of tissue-specific microRNAs from mouse. *Current Biology* 12: 735.

56. Chang, J., Nicolas, E., Marks, D., Sander, C., Lerro, A., Buendia, M. A., Xu, C., Mason, W. S., Moloshok, T., Bort, R., Zaret, K. S., and Taylor, J. M. 2004. miR-122, a mammalian liver-specific microRNA, is processed from hcr mRNA and may downregulate the high affinity cationic amino acid transporter CAT-1. *RNA Biol.* 1: 106.

57. Esau, C., Davis, S., Murray, S. F., Yu, X. X., Pandey, S. K., Pear, M., Watts, L., Booten, S. L., Graham, M., McKay, R., Subramaniam, A., Propp, S., Lollo, B. A., Freier, S., Bennett, C. F., Bhanot, S., and Monia, B. P. 2006. miR-122 regulation of lipid metabolism revealed by in vivo antisense targeting. *Cell. Metabolism* 3: 87.

58. Poy, M. N., Eliasson, L., Krutzfeldt, J., Kuwajima, S., Ma, X., MacDonald, P. E., Pfeffer, S., Tuschl, T., Rajewsky, N., Rorsman, P., and Stoffel, M. 2004. A pancreatic islet-specific microRNA regulates insulin secretion. *Nature* 432: 226.

59. Calin, G. A., Dumitru, C. D., Shimizu, M., Bichi, R., Zupo, S., Noch, E., Aldler, H., Rattan, S., Keating, M., Rai, K., Rassenti, L., Kipps, T., Negrini, M., Bullrich, F., and Croce, C. M. 2002. Frequent deletions and down-regulation of microRNA genes miR15 and miR16 at 13q14 in chronic lymphocytic leukemia. *Proc. Natl. Acad. Sci. USA* 99: 15524.

60. Johnson, S. M., Grosshans, H., Shingara, J., Byrom, M., Jarvis, R., Cheng, A., Labourier, E., Reinert, K. L., Brown, D., and Slack, F. J. 2005. RAS is regulated by the let-7 microRNA family. *Cell* 120: 635.

61. Metzler, M., Wilda, M., Busch, K., Viehmann, S., and Borkhardt, A. 2004. High expression of precursor microRNA-155/BIC RNA in children with Burkitt lymphoma. *Genes Chromosomes Cancer* 39: 167.

62. Chan, J. A., Krichevsky, A. M., and Kosik, K. S. 2005. MicroRNA-21 is an antiapoptotic factor in human glioblastoma cells. *Cancer Res.* 65: 6029.

63. Calin, G. A., and Croce, C. M. 2006. MicroRNA-cancer connection: the beginning of a new tale. *Cancer Res.* 66: 7390.

64. Monticelli, S., Ansel, K. M., Xiao, C., Socci, N. D., Krichevsky, A. M., Thai, T. H., Rajewsky, N., Marks, D. S., Sander, C., Rajewsky, K., Rao, A., and Kosik, K. S. 2005. MicroRNA profiling of the murine hematopoietic system. *Genome Biol.* 6: R71.

65. Cobb, B. S., Nesterova, T. B., Thompson, E., Hertweck, A., O'Connor, E., Godwin, J., Wilson, C. B., Brockdorff, N., Fisher, A. G., Smale, S. T., and Merkenschlager, M. 2005. T cell lineage choice and differentiation in the absence of the RNase III enzyme Dicer. *J. Exp. Med.* 201: 1367.

66. Cobb, B. S., Hertweck, A., Smith, J., O'Connor, E., Graf, D., Cook, T., Smale, S. T., Sakaguchi, S., Livesey, F. J., Fisher, A. G., and Merkenschlager, M. 2006. A role for Dicer in immune regulation. *J. Exp. Med.* 203: 2519.

67. Taganov, K. D., Boldin, M. P., Chang, K.-J., and Baltimore, D. 2006. NF-kappaB-dependent induction of microRNA miR-146, an inhibitor targeted to signaling proteins of innate immune responses. *Proc. Natl. Acad. Sci. USA* 103: 12481.

68. O'Connell, R. M., Taganov, K. D., Boldin, M. P., Cheng, G., and Baltimore, D. 2007. MicroRNA-155 is induced during the macrophage inflammatory response. *Proc. Natl. Acad. Sci. USA* 104: 1604.

69. Lecellier, C.-H., Dunoyer, P., Arar, K., Lehmann-Che, J., Eyquem, S., Himber, C., Saib, A., and Voinnet, O. 2005. A cellular microRNA mediates antiviral defense in human cells. *Science* 308: 557.

70. Jopling, C. L., Yi, M., Lancaster, A. M., Lemon, S. M., and Sarnow, P. 2005. Modulation of hepatitis C virus RNA abundance by a liver-specific microRNA. *Science* 309: 1577.

71. Hariharan, M., Scaria, V., Pillai, B., and Brahmachari, S. K. 2005. Targets for human encoded microRNAs in HIV genes. *Biochem. Biophys. Res. Commun.* 337: 1214.

72. Scaria, V., Hariharan, M., Maiti, S., Pillai, B., and Brahmachari, S. K. 2006. Host-virus interaction: a new role for microRNAs. *Retrovirology* 3: 68.

73. Cullen, B. R. 2006. Viruses and microRNAs. *Nat. Genet.* 38 (Suppl.): S25.
74. Cullen, B. R. 2006. Is RNA interference involved in intrinsic antiviral immunity in mammals? *Nat. Immunol.* 7: 563.
75. Bost, K. L. 2004. Tachykinin-modulated anti-viral responses. *Front. Biosci.* 9: 1994.
76. Bowie, A. G., and Haga, I. R. 2005. The role of Toll-like receptors in the host response to viruses. *Mol. Immunol.* 42: 859.
77. Perry, A. K., Chen, G., Zheng, D., Tang, H., and Cheng, G. 2005. The host type I interferon response to viral and bacterial infections. *Cell. Res.* 15: 407.
78. Barik, S., and Bitko, V. 2006. Prospects of RNA interference therapy in respiratory viral diseases: update 2006. *Expert Opin. Biol. Ther.* 6: 1151.
79. Dykxhoorn, D. M., and Lieberman, J. 2006. Silencing viral infection. *PLoS Med.* 3: e242.
80. Stram, Y., and Kuzntzova, L. 2006. Inhibition of viruses by RNA interference. *Virus Genes* 32: 299–306.
81. Pfeffer, S., Sewer, A., Lagos-Quintana, M., Sheridan, R., Sander, C., Grasser, F. A., van Dyk, L. F., Ho, C. K., Shuman, S., Chien, M., Russo, J. J., Ju, J., Randall, G., Lindenbach, B. D., Rice, C. M., Simon, V., Ho, D. D., Zavolan, M., and Tuschl, T. 2005. Identification of microRNAs of the herpesvirus family. *Nat. Methods* 2 (4): 269.
82. Bennasser, Y., Le, S. Y., Yeung, M. L., and Jeang, K. T. 2004. HIV-1 encoded candidate microRNAs and their cellular targets. *Retrovirology* 1: 43.
83. Andl, T., Murchison, E. P., Liu, F., Zhang, Y., Yunta-Gonzalez, M., Tobias, J. W., Andl, C. D., Seykora, J. T., Hannon, G. J., and Millar, S. E. 2006. The miRNA-processing enzyme dicer is essential for the morphogenesis and maintenance of hair follicles. *Curr. Biol.* 16: 1041.
84. Valencia-Sanchez, M. A., Liu, J., Hannon, G. J., and Parker, R. 2006. Control of translation and mRNA degradation by miRNAs and siRNAs. *Genes Dev.* 20: 515.
85. Banerjee, A. K., Barik, S., and De, B. P. 1991. Gene expression of nonsegmented negative strand RNA viruses. *Pharmacol. Ther.* 51: 47.
86. Kingsbury, D. W. 1972. Paramyxovirus replication. *Curr. Top. Microbiol. Immunol.* 59: 1.
87. Jackson, R. J., and Standart, N. 2007. How do microRNAs regulate gene expression? *Sci. STKE* 367: RE 1.
88. John, B., Sander, C., and Marks, D. S. 2006. Prediction of human microRNA targets. *Methods Mol. Biol.* 342: 101.
89. Chekanova, J. A., and Belostotsky, D. A. 2006. MicroRNAs and messenger RNA turnover. *Methods Mol. Biol.* 342: 73.
90. Marcello, A. 2006. Latency: the hidden HIV-1 challenge. *Retrovirology* 3: 7.
91. Lassen, K., Han, Y., Zhou, Y., Siliciano, J., and Siliciano, R. F. 2004. The multifactorial nature of HIV-1 latency. *Trends Mol. Med.* 10: 525.
92. Stevenson, M., Bukrinsky, M., and Haggerty, S. 1992. HIV-1 replication and potential targets for intervention. *AIDS Res. Hum. Retroviruses* 8: 107.
93. Haase, A. D., Jaskiewicz, L., Zhang, H., Laine, S., Sack, R., Gatignol, A., and Filipowicz, W. 2005. TRBP, a regulator of cellular PKR and HIV-1 virus expression, interacts with Dicer and functions in RNA silencing. *EMBO Rep.* 6: 961.
94. Boden, D., Pusch, O., Silbermann, R., Lee, F., Tucker, L., and Ramratnam, B. 2004. Enhanced gene silencing of HIV-1 specific siRNA using microRNA designed hairpins. *Nucleic Acids Res.* 32: 1154.
95. Omoto, S., and Fujii, Y. R. 2006. Cloning and detection of HIV-1-encoded microRNA. *Methods Mol. Biol.* 342: 255.
96. Omoto, S., and Fujii, Y. R. 2005. Regulation of human immunodeficiency virus 1 transcription by nef microRNA. *J. Gen. Virol.* 86: 751.

97. Omoto, S., Ito, M., Tsutsumi, Y., Ichikawa, Y., Okuyama, H., Brisibe, E. A., Saksena, N. K., and Fujii, Y. R. 2004. HIV-1 nef suppression by virally encoded microRNA. *Retrovirology* 1: 44.
98. Berkhout, B., and Haasnoot, J. 2006. The interplay between virus infection and the cellular RNA interference machinery. *FEBS Lett.* 580: 2896.
99. Alcami, A. 2003. Viral mimicry of cytokines, chemokines and their receptors. *Nat. Rev. Immunol.* 3: 36.
100. Grandvaux, N., tenOever, B. R., Servant, M. J., and Hiscott, J. 2002. The interferon antiviral response: from viral invasion to evasion. *Curr. Opin. Infect. Dis.* 15: 259.
101. Tripp, R. A., Oshansky, C., and Alvarez, R. 2005. Cytokines and respiratory syncytial virus infection. *Proc. Am. Thorac. Soc.* 2: 147.
102. Tripp, R. A. 2004. Pathogenesis of respiratory syncytial virus infection. *Viral Immunol.* 17: 165.
103. Chen, P. Y., and Meister, G. 2005. MicroRNA-guided posttranscriptional gene regulation. *Biol. Chem.* 386: 1205.
104. Wienholds, E., and Plasterk, R. H. 2005. MicroRNA function in animal development. *FEBS Lett.* 579: 5911.
105. Carrington, J. C., and Ambros, V. 2003. Role of microRNAs in plant and animal development. *Science* 301: 336.
106. Zhang, B., Pan, X., Wang, Q., Cobb, G. P., and Anderson, T. A. 2006. Computational identification of microRNAs and their targets. *Computational Biol. Chem.* 30: 395.
107. Cullen, B. R. 2002. RNA interference: antiviral defense and genetic tool. *Nat. Immunol.* 3: 597.
108. Xue, C., Li, F., He, T., Liu, G. P., Li, Y., and Zhang, X. 2005. Classification of real and pseudo microRNA precursors using local structure-sequence features and support vector machine. *BMC Bioinformatics* 6: 310.
109. Hill, A. E., Hong, J. S., Wen, H., Teng, L., McPherson, D. T., McPherson, S. A., Levasseur, D. N., and Sorscher, E. J. 2006. Micro-RNA-like effects of complete intronic sequences. *Front. Biosci.* 11: 1998.
110. Altuvia, Y., Landgraf, P., Lithwick, G., Elefant, N., Pfeffer, S., Aravin, A., Brownstein, M. J., Tuschl, T., and Margalit, H. 2005. Clustering and conservation patterns of human microRNAs. *Nucleic Acids Res.* 33: 2697.
111. Legendre, M., Lambert, A., and Gautheret, D. 2005. Profile-based detection of microRNA precursors in animal genomes. *Bioinformatics* 21: 841.
112. Poland, G. A., Jacobson, R. M., and Targonski, P. V. 2007. Avian and pandemic influenza: an overview. *Vaccine* 25: 3057.
113. Luo, G., and Palese, P. 1992. Genetic analysis of influenza virus. *Curr. Opin. Genet. Dev.* 2: 77.
114. Palese, P., Zheng, H., Engelhardt, O. G., Pleschka, S., and Garcia-Sastre, A. 1996. Negative-strand RNA viruses: genetic engineering and applications. *Proc. Natl. Acad. Sci. USA* 93: 11354.
115. Chenna, R., Sugawara, H., Koike, T., Lopez, R., Gibson, T. J., Higgins, D. G., and Thompson, J. D. 2003. Multiple sequence alignment with the Clustal series of programs. *Nucleic Acids Res.* 31: 3497.
116. Gotoh, B., Komatsu, T., Takeuchi, K., and Yokoo, J. 2001. Paramyxovirus accessory proteins as interferon antagonists. *Microbiol. Immunol.* 45: 787.
117. Hacking, D., and Hull, J. 2002. Respiratory syncytial virus: viral biology and the host response. *J. Infect.* 45: 18.
118. Johnson, T. R., Teng, M. N., Collins, P. L., and Graham, B. S. 2004. Respiratory syncytial virus (RSV) G glycoprotein is not necessary for vaccine-enhanced disease induced by immunization with formalin-inactivated RSV. *J. Virol.* 78: 6024.

119. Spann, K. M., Tran, K. C., and Collins, P. L. 2005. Effects of nonstructural proteins NS1 and NS2 of human respiratory syncytial virus on interferon regulatory factor 3, NF-kappaB, and proinflammatory cytokines. *J. Virol.* 79: 5353.
120. Ng, K. L., and Mishra, S. K. 2007. De novo SVM classification of precursor microRNAs from genomic pseudo hairpins using global and intrinsic folding measures. *Bioinformatics* 23: 1321.
121. Kasschau, K. D., Xie, Z., Allen, E., Llave, C., Chapman, E. J., Krizan, K. A., and Carrington, J. C. 2003. P1/HC-Pro, a viral suppressor of RNA silencing, interferes with Arabidopsis development and miRNA function. *Dev. Cell* 4: 205.
122. Dunn, W., Trang, P., Zhong, Q., Yang, E., van Belle, C., and Liu, F. 2005. Human cytomegalovirus expresses novel microRNAs during productive viral infection. *Cell. Microbiol.* 7: 1684.
123. Krutzfeldt, J., Kuwajima, S., Braich, R., Rajeev, K. G., Pena, J., Tuschl, T., Manoharan, M., and Stoffel, M. 2007. Specificity, duplex degradation and subcellular localization of antagomirs. *Nucleic Acids Res.* 35(9): 2885–2892.
124. Adams, B.D., Furneaux, H., and White, B. 2007. The micro-ribonucleic acid (miRNA) miR-206 targets the human estrogen receptor-alpha (ERalpha) and represses ERalpha messenger RNA and protein expression in breast cancer cell lines. *Mol. Endocrinol.* 21: 1132.
125. Qi, P., Han, J. X., Lu, Y. Q., Wang, C., and Bu, F. F. 2006. Virus-encoded microRNAs: future therapeutic targets? *Cell. Mol. Immunol.* 3: 411.
126. Krutzfeldt, J., Rajewsky, N., Braich, R., Rajeev, K. G., Tuschl, T., Manoharan, M., and Stoffel, M. 2005. Silencing of microRNAs in vivo with "antagomirs." *Nature* 438: 685.

3 Development of Gene-Profile-Responsive Antisense Agents

Sergei A. Kazakov
Somagenics, Inc.

Brian H. Johnston
Somagenics, Inc., and Department of Pediatrics,
Stanford University School of Medicine

CONTENTS

3.1 Introduction: Making Sense of Pharmacogenomics through Antisense Technologies..29
3.2 Antisense Approaches to Gene Silencing 30
3.3 Molecular Basis of Sequence Specificity of Antisense Agents 32
3.4 RNA Lassos® 37
3.5 Identifying Potent and Specific Antisense Target Sequences......................40
3.6 Conclusion..43
References 43

3.1 INTRODUCTION: MAKING SENSE OF PHARMACOGENOMICS THROUGH ANTISENSE TECHNOLOGIES

The immediate aim of pharmacogenomics is to identify genes that determine differences in individual responses to particular drugs. The longer-term goal of this emerging discipline is to advance beyond the current approach to drug therapy to more individualized approaches. Drugs that are more suited to the molecular characteristics of individual patients should have greater efficacy and reduced toxicity [1–3]. Although individual human genomes are 99.9% identical, the 0.1% difference predicts as many as 3 million polymorphisms, including substitutions, deletions, and insertions [4]. Some of these polymorphisms affect protein expression or function, and may lead to disease or altered drug response. Newly available compilations of human genome sequence polymorphisms, particularly single-nucleotide polymorphisms (SNPs), can provide markers associated with characteristic genotypes and potentially identify genes directly responsible for the individual drug responses [5–10]. Initial optimism about the ability of pharmacogenomics to speed the development of personalized medicine has been tempered by slower-than-expected progress, especially in drug target discovery and validation. One of the reasons is that drugs very rarely interact with only a single target, and even when they do, they usually affect several pathways [11, 12]. Even fewer drugs can distinguish between protein variants that differ as a result of minor polymorphisms such as SNPs. Many drug development programs still employ the traditional approach of identifying a single target and using combinatorial chemistry and high-throughput assays to identify drug leads. However, companies are increasingly employing a more directed approach using information provided by pharmacogenomics and structural biology.

Nucleic-acid-based antisense technologies are particularly amenable to rational design because of the straightforward pairing relationship between the sequences of the antisense agent and its RNA target. Moreover, the availability of the complete human genome sequence has given the antisense approach a new and powerful resource. While antisense technologies are conceptually elegant and straightforward, in practice their specificity and potency *in vivo* are unpredictable. In this chapter, we focus on the ability of various antisense agents to distinguish sequence polymorphisms and to access their intended target sites. These capabilities are central to their utility as gene-profile-responsive therapeutics. Other important issues include the efficient delivery of antisense agents into the appropriate cellular compartment, and their biostability (not covered by this chapter). Small interfering RNA (siRNA) has been extensively reviewed elsewhere, so the focus of this chapter is primarily on other antisense approaches.

3.2   ANTISENSE APPROACHES TO GENE SILENCING

Although antisense technologies have been under development for more than 25 years [13–22], antisense-mediated gene regulation has more recently been found to occur in a variety of natural systems, in the forms of antisense RNA [23–26], ribozymes [27–29], and RNA interference [30–33]. From a mechanistic point of view, there are five classes of nucleic-acid-based agents that can mediate gene knockdown via Watson-Crick pairing, which is the defining feature of antisense recognition. The mechanistically simplest class is synthetic antisense oligonucleotides that rely on covalent or strong noncovalent binding to the target RNA, resulting in steric blockage of translation without cleaving the target. Noncovalent blockers include nucleic acid analogs such as phosphorodiamidate morpholino oligomers (PMOs), N3'→P5' phosphoramidates, 2'-O-methoxyethyl RNAs, locked nucleic acids (LNA), and peptide nucleic acids (PNA) [34–37]. Covalent-bond blocking agents comprise antisense oligonucleotides with reactive groups, such as alkylating, platinum, and photoactive derivatives [38–40], that are capable of covalent cross-linking with their RNA targets after hybridization.

A second class includes synthetic antisense oligodeoxynucleotides (ODNs), as well as certain nuclease-resistant analogs such as phosphorothioate or 2'-fluoroarabino derivatives that can recruit the endogenous ribonuclease RNase H to cleave a target RNA upon hybridizing with it [41–42]. Although partially phosphothioate-modified oligonucleotides were the basis for a wide variety of "first-generation" antisense drug candidates, only one has been approved by the FDA: Vitravene (fomivirsen), which is used against cytomegalovirus infections of the eye, primarily in AIDS patients. In general, first-generation chemistries are less potent and have worse side-effects than subsequent designs [43]. More successful have been so-called second-generation chemistries, which combine phosphorothioate modifications, allowing RNase H cleavage, with 2'-ribose modifications of residues near the ends of the oligonucleotide that provide increased helix stability, such as 2'-O-methyl and 2'-O-methoxyethyl [44]. The stringency requirements of RNase H are low, and as little as a 5–7 bp of complementarity between such oligonucleotides and an intracellular RNA molecule may be sufficient to cleave the RNA, leading to unintended effects [41, 45, 46]. Strategies using alternative endogenous ribonucleases that avoid this issue include conjugation with 2',5'-oligoadenylate, which recruits RNase L [47, 48], or mimicking the 3'-end of tRNA to recruit RNase P [49–51].

A third class is made up of catalytic nucleic acids, ribozymes, and deoxyribozymes that can hybridize to and cleave target RNAs without the need for any protein cofactor such as RNase H [17, 52–54]. A variant of this class, a group sometimes called artificial ribonucleases, consists of ODNs bearing catalytic groups that can cleave RNA either directly or with the assistance of metal ions [55–58].

A fourth class includes antisense RNAs expressed intracellularly from appropriate vectors [59–68]. Although such RNAs inherently lack chemical stabilization against nuclease degradation, they can be highly potent inhibitors of gene expression [69–72]. Antisense RNAs pair with RNA targets with a binding strength similar to that of the steric blockers (N3'→P5' phosphoramidates, morpholino phosphorodiamidates, and 2'-O-methoxyethyl modifications) [36, 37, 73]. Antisense RNA can inactivate its target mRNA by the physical blocking or disruption of functionally active structures (for example, preventing splicing, export from nucleus to cytoplasm, or initiation of translation), or by induction of target RNA degradation by cellular nucleases such as RNAse III [23, 74–80].

The final group is small interfering RNA, which, unlike other antisense classes, are duplex rather than single-stranded oligonucleotides. Three subgroups can be distinguished, each typically 19–29 bp in length: small interfering RNAs (siRNAs), small hairpin RNAs (shRNAs), and the natural cousins of shRNAs, microRNAs (miRNAs) [81–83]. In the case of siRNAs, the RNA duplex becomes incorporated into an RNA-induced silencing complex (RISC), one of the two strands (the sense strand) is displaced, and subsequent pairing of the antisense strand with the target mRNA leads to cleavage of the latter at the binding site, catalyzed by the Argonaute 2 component of RISC. Moreover, recent results concerning the complexes formed between antisense RNA and target RNA provide direct evidence for mechanistic links between antisense-mediated gene silencing and posttranscriptional gene silencing through RNA interference and suggest that their mechanisms of action could share steps in common at the double-stranded RNA stage [84–87].

siRNAs (including shRNAs), which were first developed only a few years ago, are generally more potent than other classes of antisense agents [88, 89]. However, there are currently several challenges to the use of siRNAs. Some RNAs appear to be poor targets for siRNAs. These may include highly structured RNAs [90–93], short-lived transcripts, and viral RNAs having high mutation rates [88, 94]. Many RNA viruses, whose hosts include plants as well as mammals, have found ways to escape from the inhibitory effects of siRNAs by antagonizing RNA interference pathways [95–105]. siRNAs can also have significant "off-target" effects, knocking down unintended mRNAs that have similar sequences [106–114]. In addition to affecting specific "off-targets," several recent reports indicate that when siRNAs are introduced into cells, they can induce a nonspecific interferon response [115–120]. Finally, high-level expression of shRNA can be toxic, perhaps due to saturation of the cellular RNAi pathways that are needed for miRNA function [121].

Thus, traditional antisense methods may still provide useful alternative or complementary approaches for gene knockdown when siRNAs are insufficiently effective or specific [17, 122–126]. Indeed, "morpholino" oligonucleotides are as potent as siRNAs in inhibiting the translation of certain mRNA targets [73, 127, 128]. The same is true of some other types of antisense oligonucleotides whose sequences have been carefully optimized [15, 16, 36, 43, 129–134]. It is also to be noted that treatment of certain conditions requires calibrated modulation of a target mRNA rather than its complete inhibition [135].

Besides preventing translation, some antisense agents (those that block rather than cleave their targets) can be used for regulation of alternative splicing [136–143] and intracellular imaging of gene expressions [144–148]. It is important to emphasize that siRNAs are not suitable for these purposes. Yet another application, one that is currently generating particular excitement, and for which antisense oligonucleotides are uniquely suited, is the knockdown of miRNAs through the formation of stable, inactive antisense-miRNA duplexes [149–155]. Since some miRNAs are implicated in cancer [9, 156–158] and viral infection [159–161], antisense agents directed against these miRNAs are potential drug candidates.

3.3 MOLECULAR BASIS OF SEQUENCE SPECIFICITY OF ANTISENSE AGENTS

Antisense-based drugs and hybridization probes (we use the term *antisense agents* to describe both) share a very important feature: They are all designed to sequence-specifically bind their polynucleotide targets through complementary (usually Watson-Crick) base-pairing. They can also bind to imperfectly complementary (mismatched) sequences, but with a reduced affinity compared with perfectly matched partners. The differences in thermostability between a perfect duplex and a mismatched duplex depend on length, GC content and sequence, as well as the type and position of mismatches. In the presence of a limited number of different targets *in vitro*, shorter antisense agents are generally more selective than longer ones [40, 162–165]. However, when many different sequences are simultaneously present (such as in microarrays or cells), the probability of finding short, perfectly matching sequences in both target and nontarget polynucleotides increases, limiting the overall selectivity of short antisense sequences.

Assuming a distribution of nucleotides in the human genome, whose estimated length is 3×10^9 bp [166], antisense sequences longer than 18 nt would be required to ensure complete uniqueness of the perfectly matched complexes with the genome sequences. However, only an estimated 2–3% (~1×10^8 nt) of the genome encodes mature into mRNA sequences [167]. Because only about 25% of these mRNAs are thought to be expressed in any given cell, the actual number of targets is closer to 2×10^7 nt [168]. Of this number, less than 20% are estimated to be accessible because of formation of stable secondary structures and RNA-protein complexes [40, 168, 169]. Thus, only about 4×10^6 nt are targetable, and hence an antisense molecule of as few as 11 nt ($4^{11} = 4.2 \times 10^6$ nt) could be long enough to uniquely match a specific mRNA target sequence. (Note that 11 bp is about the length of one turn of the A-form helix formed by both RNA-RNA and DNA-RNA duplexes.) Interestingly, several studies report that oligonucleotides only 7–11 nt in length can selectively inhibit translation of certain genes through targeting their mRNAs [170–175], although long oligonucleotides are usually far more potent inhibitors of gene expression than shorter ones [22, 40].

In general, RNA-RNA hybrids are more stable than the corresponding DNA-RNA and DNA-DNA duplexes [176–179]. As indicated above, the efficiency of probe-target hybridization can be hindered by the formation of secondary structures that reduce target site accessibility [180, 181]. For structured targets, longer probes (>20 nt) usually are more effective and have higher affinity to RNA targets than shorter probes because they have multiple opportunities for base-pairing. Because of the generally higher duplex stability of RNA-RNA pairing over DNA-RNA, antisense RNA probes have faster hybridization kinetics and a better ability to bind structured polynucleotide targets than corresponding DNA probes [165, 182]. It appears that high affinity for their target sites and fast hybridization kinetics [183–187] are the most important determinants of efficacy in the case of antisense agents that act through noncleaving, target-blocking mechanisms.

The trade-offs between high affinity for the target and low sequence specificity of binding have implications for the design of allele-specific antisense agents [40, 188]. Increasing the affinity of antisense agents to their intended polynucleotide targets, either by lengthening antisense sequences or by using appropriate chemical modifications of antisense oligonucleotides, will simultaneously decrease their selectivity, thereby enhancing off-target effects [40]. This limitation is also one of the major hurdles in the development of antisense agents that can discriminate SNPs and hence distinguish individual gene profiles. This is true even for *in vitro* assays that can otherwise be optimized for maximum selectivity (e.g., through variation of temperature, incubation time, salt and formamide concentration of the hybridization, and washing conditions). Antisense drugs lack this option because the intracellular environment provides fixed conditions, including constant physiological temperature and solutes.

There are several ways to design SNP-sensitive antisense agents. The first approach is to use antisense chemistries that provide tight binding even for short pairing regions. In this way, a single mismatch has a large impact on the helical stability, yet a short sequence is stable at body temperature. In the case of LNA, each substitution of an LNA residue for a DNA increases the melting temperature (T_m) by 2°C–10°C per LNA monomer (depending on sequence content) when hybridized to RNA targets [189, 190]. LNAs have been successfully used for this purpose [191], as have morpholinos [128].

Another approach uses side-by-side hybridizing oligonucleotides that can cooperatively bind adjacent sites in RNA targets [192, 193] (Figure 3.1).

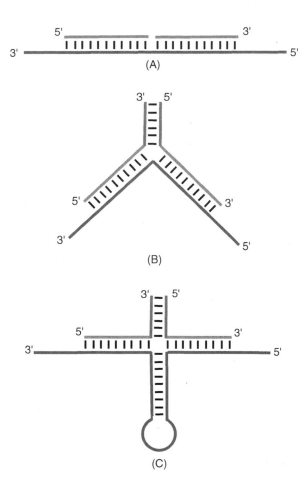

FIGURE 3.1 (See color insert following page 56.) Cooperative binding of two short oligonucleotides to RNA targets. (A) Side-by-side binding of two oligonucleotides to adjacent target sequences. The complex is stabilized through stacking interactions at the interface between the oligonucleotides. (B) Side-by-side binding of two partially complementary oligonucleotides to adjacent target sequences. The complex is stabilized through base-pairing between the oligonucleotide dimerization segments. (C) Binding of two partially complementary oligonucleotides to nonadjacent target sequences that are brought together in space by a secondary structure in the target. This complex is also stabilized through base-pairing between the oligonucleotide dimerization segments. RNA targets are shown in blue, antisense in red, and the dimerization segments in green.

A third approach is to use ribozymes that employ two comparatively short antisense sequences (binding arms) acting in concert, such as hammerhead (7 + 7 nt) [194, 195] (Figure 3.2A) and hairpin ribozymes (6 + 4 nt) [194, 196] (Figure 3.2B). Indeed, SNP specificity of RNA cleavage by hammerhead ribozymes has been demonstrated *in vitro* [197, 198]. However, other results indicate that hammerhead ribozymes with longer arms, which are less sensitive to mismatches, are substantially more potent in cells than those with short arms [199–201]. Short antisense RNAs

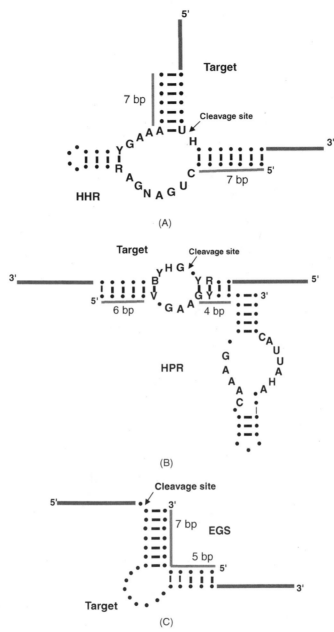

FIGURE 3.2 (See color insert following page 56.) Secondary structures and consensus sequences of representative ribozymes cleaving their RNA targets in bimolecular reactions (*in trans*). (A) Hammerhead ribozyme (HHR). (B) Hairpin ribozyme (HPR). (C) External guide sequence (EGS) directing cleavage of target RNA by the human RNase P ribozyme. Dots represent any nucleotide (A, U, G, or C); dashes represent required pairings; V is "not U" (A, C, or G); Y is a pyrimidine (U or C); R is a purine (A or G); B is "not A" (U, C, or G); and H is "not G" (A, C, or U) [275]. RNA targets are shown in blue, and antisense ribozyme arms are in red.

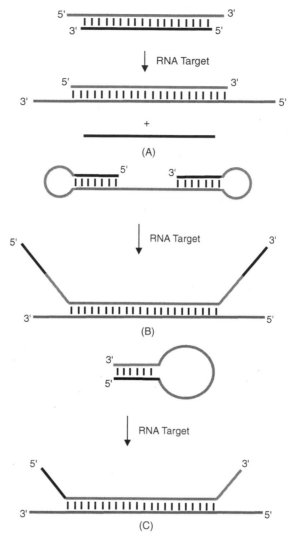

FIGURE 3.3 (See color insert following page 56.) Antisense oligonucleotides equipped with various types of stringency elements. (A) Antisense oligonucleotide prehybridized with a complementary masking oligonucleotide that covers the target site but is shorter by a few nucleotides at one or both ends. As a result of the competitive hybridization, the antisense sequence forms a perfect duplex with the target, and the masking oligonucleotide gets displaced. (B) Antisense sequence extended at either one or both ends (two-end extension is shown) by sequences forming terminal hairpin structures. As a result of the competitive hybridization with the target, the antisense sequence forms a perfect duplex, whereas the terminal masking sequences gets displaced. (C) Antisense sequence extended at both ends by short complementary sequences that form a stem-and-loop structure known as a "molecular beacon." When the antisense sequence in the loop anneals to a complementary target sequence, the longer and stronger probe-target duplex overcomes the internal secondary structure, leading to opening of this structure. Antisense oligonucleotides having all these stringency elements form stable, perfect duplexes with normal target sequences, whereas targets containing mismatches form either unstable duplexes, or no duplexes. RNA targets are in blue, antisense in red, and the stringency elements are in black and green as shown.

encoding external guide sequences for human RNase P, consisting of two short (7 + 5 nt) arms [202] (Figure 3.2C), can induce cleavage of the target RNA in cells while providing true SNP sensitivity [203].

A fourth approach is the use of stringency elements that can improve mismatch discrimination upon hybridization, including displacement hybridization [204, 205] (Figure 3.3A), hairpins [206–208] (Figure 3.3B), and molecular beacons [209, 210] (Figure 3.3C).

A fifth approach is the introduction of artificial mismatches in antisense sequences [211, 212] (Figure 3.4). Because two mismatches a certain distance apart are especially destabilizing to a duplex, introducing an artificial mismatch that distance from an SNP is an effective way to increase the discrimination between the SNP variants. This last approach is similar to what is often found in natural antisense mechanisms [23–24, 212–214]. Most naturally occurring regulatory antisense RNAs, which are typically 60–100 nt in length and transcribed from a locus different from that encoding the target RNA, do not perfectly match their RNA targets [23, 212]. Complexes of natural ribozymes and aptamers with their targets also frequently contain mismatches and noncanonical base-pairing [52, 54, 215, 216), suggesting that such structural elements may function to enhance sequence specificity.

3.4   RNA LASSOS®

An ideal antisense agent, capable of SNP discrimination, would combine the excellent hybridization efficacy and high target affinity of long antisense sequences with the ability of short antisense sequences to discriminate closely related gene sequences. To avoid the usual trade-off between these two desirable features, one might try to use short recognition sequences to assure high specificity while stabilizing the antisense-target duplex by other means once it is formed. One manifestation of this approach is the RNA Lasso [217–221] (Figure 3.5). RNA Lassos contain an internal hairpin ribozyme (HPR) moiety (Figure 3.5A) that has both self-cleavage and self-ligating capabilities [196]. This HPR moiety exists as a dynamic equilibrium between linear and circular forms that can be regulated by features of the ribozyme sequence [196]. The size and sequence of the three loops that connect the helical segments of the HPR can be modified without substantial effects on catalytic activity if the loop sequences do not interfere with the proper folding of the ribozyme [222–224]. The principle of the Lasso is that insertion of an antisense sequence into one of these loops (typically loop 2) allows the Lasso to pair with a target mRNA, intertwining the two RNAs if it is in the linear form. The Lasso can then self-circularize, creating a "topologically linked" complex with a linear target mRNA. (If the target were itself circular, the complex would have a true topological linkage; in the case of a linear target, the complex could dissociate only if all pairing with the antisense sequence were disrupted and the Lasso slipped off the end, or if the Lasso were to become linear through cleavage of its backbone.) In contrast to the conventional application of ribozymes as sequence-specific nucleases, Lassos do not cleave their targets, but by linking to them, they create complexes that are thermodynamically more stable than ordinary RNA-RNA duplexes. Another advantage of circularity is that it makes Lassos resistant to exonucleases.

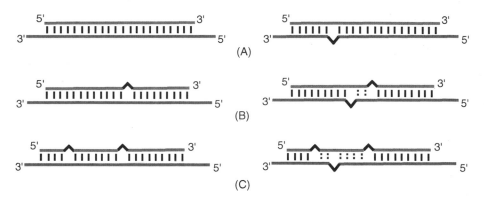

FIGURE 3.4 (See color insert following page 56.) Perfect and mismatched duplexes between an antisense oligonucleotide and an RNA target. (A) Conventional allele-specific hybridization of a "perfect" antisense with either a normal target (left) or one with a single-nucleotide substitution due to a mutation or SNP (right). (B) Hybridization of a single-base mismatched oligonucleotide with the normal (left) and mutated targets (right). (C) Hybridization of doubly mismatched oligonucleotide with normal (left) and mutated targets (right). In all cases the oligonucleotide forms a more stable duplex with the normal target than the mutated target; however, because mismatches spaced a certain distance apart are especially destabilizing, the presence of two or three mismatches between antisense oligonucleotide and the target may provide better discrimination between the two targets. RNA targets are shown in blue, antisense in red, and mismatches in black. The interactions between complementary bases that are weakened by the nearby mismatches are shown as dotted lines.

Lassos can be highly sequence-specific [221], presumably due at least in part to their circularity. It has been shown that antisense DNA circles show higher sequence selectivity (more destabilization by the presence of mismatches) than do linear ODNs of the same sequence [225, 226]. Because stable binding requires disruption of some secondary structure (Loop A and Helix 2 in Figure 3.5A), the latter may act as a stringency element (see above) by competing with the target for binding and thereby reducing the net energy gain upon binding. When binding is weaker at a given temperature, the complex is more prone to destabilization by mismatches. In this respect, Lassos are reminiscent of molecular beacons (see Figure 3.3C). In some cases, where a given Lasso is not optimally specific, it can be made fully SNP-sensitive by incorporating an additional stringency element (see above) into its antisense domain [221].

In their ability to topologically link around a target, RNA Lassos are also reminiscent of the Padlock probes [227–229], although they differ in several respects (Figure 3.6). Lassos are 120–130-nt RNAs that can be either transcribed *in vitro* or expressed from DNA vectors *in situ*. In contrast, Padlock probes are 70–100-nt synthetic DNAs that allow target-dependent ligation of their ends by a DNA ligase. The need for an exogenous protein ligase largely restricts the use of Padlock probes to diagnostic assays [228, 230], whereas RNA Lassos can function autonomously inside cells and are therefore of therapeutic as well as diagnostic interest.

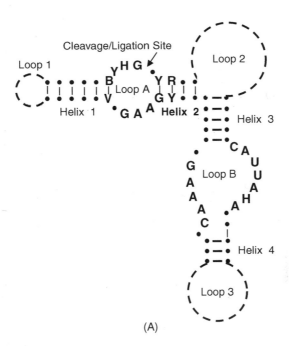

(A)

FIGURE 3.5 Structure and self-processing properties of an RNA Lasso. (A) Consensus structure of the hairpin ribozyme (HPR), derived from sequences in the minus strand of tobacco ringspot virus satellite RNA. Self-cleavage at the site shown produces 2′,3′-cyclic phosphate and 5′-OH termini. The reverse reaction can efficiently ligate those ends, and the molecule shown can exist as a dynamic equilibrium between linear and circular forms. The position of the equilibrium depends on the relative stability of the cleaved and ligated forms. Dots represent any nucleotide (A, U, G, or C); dashes represent base pairings; V is "not U" (A, C, or G); Y is a pyrimidine (U or C); R is a purine (A or G); B is "not A" (U, C, or G); and H is "not G" (A, C, or U) [275]. (B) Scheme of Lasso self-processing. Trimming the ends of a longer RNA Lasso precursor (typically made by *in vitro* transcription) through self-cleavage generates semiprocessed intermediates and the fully processed linear form, which can then convert into the circular form through self-ligation.

In addition to self-ligation, a convenient feature of the HPR is its ability to excise itself from a primary transcript, cleaving off all irrelevant flanking sequences at both the 5′ and 3′ ends, prior to circularizing (Figure 3.5B). This allows the Lasso to be independent of any sequences that may be introduced for convenience, e.g., to aid in expression. Thus, RNA Lassos can be either transcribed *in vitro* from DNA vectors by T7 RNA polymerase and delivered to cellular targets directly (e.g., in liposomal complexes), or expressed *in vivo* by RNA pol II or pol III, using appropriate plasmid, PCR-amplicon, or viral vectors.

Like antisense RNA, Lassos can potentially disable a target RNA either by physically blocking its function, causing misfolding of functionally active structures, or inducing its degradation by cellular nucleases. Which mechanism predominates can be controlled by the design features of the Lasso.

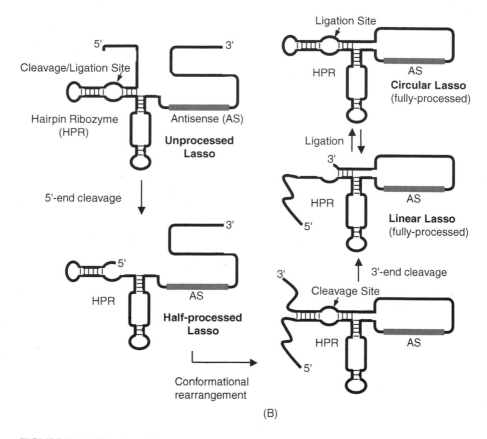

FIGURE 3.5   (Continued.)

3.5   IDENTIFYING POTENT AND SPECIFIC ANTISENSE TARGET SEQUENCES

The original concept of antisense-based drug design embodied the notion that a gene of interest can be specifically targeted as long as its sequence is known. However, several facts argue that this concept is oversimplified [133, 231–235]. First, not all sites of target RNA are equally accessible for hybridization under physiological conditions. Second, some antisense sequences may regulate multiple related genes. Third, antisense molecules can frequently form imperfect (mismatched) complexes with unintended sequences (off-target effects). Fourth, certain antisense sequences can nonspecifically alter the expression of unrelated genes, resulting in toxic effects. All of these problems make the targeting of certain sites (selected, for example, based on pharmacogenomics considerations) difficult.

There are a few approaches that in some cases can make poorly accessible sites more accessible for antisense oligonucleotides and ribozymes/deoxyribozymes, including use of longer antisense sequences (also discussed above), use of chemical modifications that increase thermostability of complementary complexes (see above), and use of helper/facilitator oligonucleotides hybridized to sequences

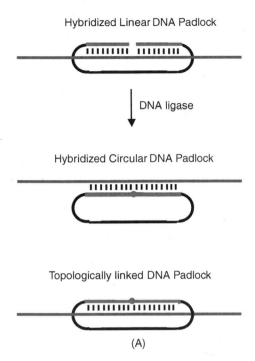

Hybridized Linear DNA Padlock

DNA ligase

Hybridized Circular DNA Padlock

Topologically linked DNA Padlock

(A)

FIGURE 3.6   (See color insert following page 56.) Circularizable nucleic acid agents. (A) Padlock Probe (DNA). (B) RNA Lasso. These agents are linear polynucleotides that can hybridize by their antisense segments (shown in red) to an RNA target (blue). Their terminal sequences are joined by either DNA ligase (Padlock Probe) or self-ligated by the encoded ribozyme (RNA Lasso). Note that the ligation site (dot) for the Padlock Probe lies within the antisense-target duplex, whereas for the RNA Lasso it is outside this duplex. Circularization of linear forms of these agents prebound to their targets results in the formation of topologically linked complexes.

flanking the target site [236–239]. However, even where accessibility is not a problem, the function of a sequence within an mRNA can also influence its effectiveness as an antisense target. MicroRNAs mainly target 3′-UTRs, while most effective antisense drugs target 5′-UTRs [128]. Noncleaving antisense agents that target the coding mRNA regions are in general less efficient, presumably because the strong helicase activity of ribosome complexes can displace even strongly bound antisense agents such as morpholino oligonucleotides [175]. This displacement activity represents another challenge for the use of antisense agents to target SNPs within coding regions.

For all of these reasons, there is a need for the development of convenient and reliable methods for identifying the most "sensitive" target sequences. Most approaches to addressing this need have involved either computer prediction or *in vitro* selection/mapping using combinatorial libraries [240–254]. Although *in vitro* screening of libraries of antisense agents by methods such as mapping mRNA targets by RNAse H cleavage or scanning target sequences by microarrays have had some success in predicting targetable sequences, many other studies show little or no correlation

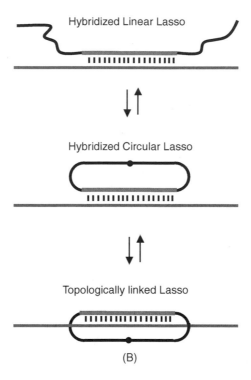

Hybridized Linear Lasso

Hybridized Circular Lasso

Topologically linked Lasso

(B)

FIGURE 3.6   (Continued.)

between the *in vitro* data and the intracellular efficacy of the same antisense agents [135, 255–259]. This incongruity may reflect different folding of the RNAs within the microenvironments of the living cell versus in the test tube, or result from the presence of RNA-binding proteins in cells. Ideally, all possible target-specific RNA sequences within an appropriate range of lengths should be prepared and individually tested in cells to ensure finding the best inhibitors for a given mRNA. However, such a "brute force" approach is expensive and time consuming.

As an alternative, screening for target sites can be performed by using random (degenerate) libraries of antisense sequences directly in cells [260–262]. However, this approach has several major problems. First, the high complexity of random libraries (e.g., 4^{20} or $\approx 10^{12}$ molecules for 20-nt antisense sequences represented only about once in the human genome) [166] may make this approach prohibitively expensive for cell-based assays. This is because cell-based selection requires either approximately one construct per cell, or if there are multiple constructs per cell, each must be potent enough to provide the basis for selection even though it is diluted by the presence of inactive members of the library. In the latter case, subsequent rounds of selection must be made to identify the active species from among the inactive ones present in the selected cells [262]. Second, since each sequence in a degenerate library has its complement also represented, the two can form stable duplexes, thus reducing their availability for interaction with accessible target sites [263]. Third, experiments have shown that degenerate libraries are highly toxic to cells because they can block the functioning of unintended cellular RNAs as well as the intended target [262, 264, 265].

Directed, or gene-specific, oligonucleotide libraries composed of all 15–25-nt long sequences represented within the target gene(s) of interest offer a superior alternative to screening completely random libraries. The use of directed libraries prepared *in vitro* significantly simplifies cell-based screening, since comparatively small libraries need to be assayed. For example, a 20-nt directed library targeting a 2000-nt mRNA consists of only 1981 distinct molecules. Moreover, unintended knockdown of nontargeted genes is reduced, allowing more efficient cell-based assays with the directed libraries cloned into appropriate vectors. Several methods for preparing directed libraries that can be cloned, amplified, and inserted into appropriate antisense, ribozyme, or siRNA expression vectors have been described [246, 264–270].

Preparation of high-quality libraries is important, but represents the simpler part of the selection procedure. The functional screening of such libraries in cell-based environments is more complicated and problematic. The development of adequate cellular screening methods is critical to identifying the most potent and least toxic antisense agents. This subject has been reviewed elsewhere [135, 270–274].

3.6 CONCLUSION

Because antisense agents obtain their specificity from Watson-Crick pairing rules, they are natural candidates for exploiting the often-subtle DNA sequence differences among individuals that are the basis of pharmacogenomics. However, because the RNA sequences affected by those polymorphisms may have a secondary structure or be bound to proteins, not all the polymorphic sites are readily accessible to antisense agents. The inability to pick and choose accessible sites is why the use of antisense for pharmacogenomics is a bigger challenge than its use for ordinary therapeutic goals, where there is more choice of target sites. Inroads have been made on this problem by two types of advances: more potent antisense technologies, including strong RNA blockers and siRNAs, and better methods for identifying good target sites so that the likelihood of finding an "antisensitive" site that overlaps a relevant polymorphism is greater. Ultimately, the role of antisense agents in this area may be as sequence-sensitive adjuncts to less specific but more potent small-molecule drugs, consistent with current trends toward the use of nucleic acids in combination therapies.

REFERENCES

1. Ginsburg, G. S., and McCarthy, J. J. 2001. Personalized medicine: revolutionizing drug discovery and patient care. *Trends Biotechnol.* 19: 491.
2. Evans, W. E., and Relling, M. V. 2004. Moving towards individualized medicine with pharmacogenomics. *Nature* 429: 464.
3. Roses, A. D. 2004. Pharmacogenetics and drug development: the path to safer and more effective drugs. *Nat. Rev. Genet.* 5: 645.
4. Hallberg, P., and Melhus, H. 2004. Candidate genes in the pharmacogenomics of antihypertensive treatment: a review and future aspects. *Curr. Pharmacogenomics* 2: 83.

5. Kirkwood, S. C., and Hockett, R. D., Jr. 2002. Pharmacogenomic biomarkers. *Dis. Markers* 18: 63.

6. Roses, A. D. 2002. SNPs: where's the beef? *Pharmacogenomics J.* 2: 277.

7. Dracopoli, N. C. 2003. Pharmacogenomic applications in clinical drug development. *Cancer Chemother. Pharmacol.* 52 (Suppl. 1): S57.

8. Roses, A. D., Burns, D. K., Chissoe, S., Middleton, L., and St. Jean, P. 2005. Disease-specific target selection: a critical first step down the right road. *Drug Discov. Today* 10: 177.

9. Hernando, E. 2007. MicroRNAs and cancer: role in tumorigenesis, patient classification and therapy. *Clin. Transl. Oncol.* 9: 155.

10. Saunders, M. A., Liang, H., and Li, W. H. 2007. Human polymorphism at microRNAs and microRNA target sites. *Proc. Natl. Acad. Sci. USA* 104: 3300.

11. Thayer, A. M. 2003. Biomarkers emerge. *Chem. Eng. News* 81: 33.

12. Mencher, S. K., and Wang, L. G. 2005. Promiscuous drugs compared to selective drugs (promiscuity can be a virtue). *BMC Clin. Pharmacol.* 5: 3.

13. Stein, C. A., and Krieg, A. M., eds. 1998. *Applied antisense oligonucleotide technology.* New York: Willey-Liss.

14. Crooke, S. T., ed. 2001. *Antisense drug technology: principles, strategies, and applications.* New York: Marcel Dekker.

15. Opalinska, J. B., and Gewirtz, A. M. 2002. Nucleic-acid therapeutics: basic principles and recent applications. *Nat. Rev. Drug Discov.* 1: 503.

16. Carmichael, G. G. 2003. Antisense starts making more sense. *Nat. Biotechnol.* 21: 371.

17. Scherer, L. J., and Rossi, J. J. 2003. Approaches for the sequence-specific knockdown of mRNA. *Nat. Biotechnol.* 21: 1457.

18. Goodchild, J. 2004. Oligonucleotide therapeutics: 25 years agrowing. *Curr. Opin. Mol. Ther.* 6: 120.

19. Ravichandran, L. V., Dean, N. M., and Marcusson, E. G. 2004. Use of antisense oligonucleotides in functional genomics and target validation. *Oligonucleotides* 14: 49.

20. Vidal, L., Blagden, S., Attard, G., and de Bono, J. 2005. Making sense of antisense. *Eur. J. Cancer* 41: 2812.

21. Chan, J. H., Lim, S., and Wong, W. S. 2006. Antisense oligonucleotides: from design to therapeutic application. *Clin. Exp. Pharmacol. Physiol.* 33: 533.

22. Pan, W. H., and Clawson, G. A. 2006. Antisense applications for biological control. *J. Cell. Biochem.* 98: 14.

23. Kumar, M., and Carmichael, G. G. 1998. Antisense RNA: function and fate of duplex RNA in cells of higher eukaryotes. *Microbiol. Mol. Biol. Rev.* 62: 1415.

24. Wagner, E. G., Altuvia, S., and Romby, P. 2002. Antisense RNAs in bacteria and their genetic elements. *Adv. Genet.* 46: 361.

25. Yelin, R., Dahary, D., Sorek, R., Levanon, E. Y., Goldstein, O., Shoshan, A., Diber, A., Biton, S., Tamir, Y., Khosravi, R., Nemzer, S., Pinner, E., Walach, S., Bernstein, J., Savitsky, K., and Rotman, G. 2003. Widespread occurrence of antisense transcription in the human genome. *Nat. Biotechnol.* 21: 379.

26. Munroe, S. H., and Zhu, J. 2006. Overlapping transcripts, double-stranded RNA and antisense regulation: a genomic perspective. *Cell. Mol. Life Sci.* 63: 2102.

27. Doherty, E. A., and Doudna, J. A. 2001. Ribozyme structures and mechanisms. *Annu. Rev. Biophys. Biomol. Struct.* 30: 457.

28. Been, M. D. 2006. HDV ribozymes. *Curr. Top. Microbiol. Immunol.* 307: 47.

29. Kazantsev, A. V., and Pace, N. R. 2006. Bacterial RNase P: a new view of an ancient enzyme. *Nat. Rev. Microbiol.* 4: 729.

30. Hannon, G. J. 2002. RNA interference. *Nature* 418: 244.

31. Baulcombe, D. 2004. RNA silencing in plants. *Nature* 431: 356.

32. Mello, C. C., and Conte, D., Jr. 2004. Revealing the world of RNA interference. *Nature* 431: 338.
33. Buchon, N., and Vaury, C. 2006. RNAi: a defensive RNA-silencing against viruses and transposable elements. *Heredity* 96: 195.
34. Micklefield, J. 2001. Backbone modification of nucleic acids: synthesis, structure and therapeutic applications. *Curr. Med. Chem.* 8: 1157.
35. Toulme, J.-J., 2001. New candidates for true antisense. *Nat. Biotechnol.* 19: 17.
36. Braasch, D. A., and Corey, D. R. 2002. Novel antisense and peptide nucleic acid strategies for controlling gene expression. *Biochemistry* 41: 4503.
37. Kurreck, J. 2003. Antisense technologies: improvement through novel chemical modifications. *Eur. J. Biochem.* 270: 1628.
38. Knorre, D. G., and Vlassov, V. V. 1991. Reactive oligonucleotide derivatives as gene-targeted biologically active compounds and affinity probes. *Genetica* 85: 53.
39. Vlassov, V. V., Vlassova, I. E., and Pautova, L. V. 1997. Oligonucleotides and poly-nucleotides as biologically active compounds. *Prog. Nucleic Acid Res. Mol. Biol.* 57: 95.
40. Toulme, J. J., Di Primo, C., and Moreau, S. 2001. Modulation of RNA function by oli-gonucleotides recognizing RNA structure. *Prog. Nucleic Acid Res. Mol. Biol.* 69: 1.
41. Dias, N., and Stein, C. A. 2002. Antisense oligonucleotides: basic concepts and mecha-nisms. *Mol Cancer Ther.* 1: 347.
42. Crooke, S. T. 2004. Antisense strategies. *Curr. Mol. Med.* 4: 465.
43. Crooke, S. T. 2004. Progress in antisense technology. *Annu. Rev. Med.* 55: 61.
44. Prakash, T. P., and Bhat, B. 2007. 2'-Modified oligonucleotides for antisense therapeu-tics. *Curr. Top. Med. Chem.* 7: 641.
45. Larrouy, B., Boiziau, C., Sproat, B., and Toulme, J. J. 1995. RNase H is responsible for the non-specific inhibition of in vitro translation by 2'-O-alkyl chimeric oligonucle-otides: high affinity or selectivity, a dilemma to design antisense oligomers. *Nucleic Acids Res.* 23: 3434.
46. Stein, C. A. 2000. Is irrelevant cleavage the price of antisense efficacy? *Pharmacol Ther.* 85: 231.
47. Adah, S. A., Bayly, S. F., Cramer, H., Silverman, R. H., and Torrence, P. F. 2001. Chem-istry and biochemistry of 2',5'-oligoadenylate-based antisense strategy. *Curr. Med. Chem.* 8: 1189.
48. Torrence, P. F., and Wang, Z. 2001. Accelerating RNA decay through intervention of RNase L: alternative synthesis of composite 2',5'-oligoadenylate-antisense. *Methods Enzymol.* 342: 20.
49. Hartmann, R. K., Krupp, G., and Hardt, W. D. 1995. Towards a new concept of gene inactivation: specific RNA cleavage by endogenous ribonuclease P. *Biotechnol. Annu. Rev.* 1: 215.
50. Guerrier-Takada, C., and Altman, S. 2000. Inactivation of gene expression using ribo-nuclease P and external guide sequences. *Meth. Enzymol.* 313: 442.
51. Trang, P., Kim, K., and Liu, F. 2004. Developing RNase P ribozymes for gene-targeting and antiviral therapy. *Cell Microbiol.* 6: 499.
52. Sioud, M., and Leirdal, M. 2000. Therapeutic RNA and DNA enzymes. *Biochem. Pharmacol.* 60: 1023.
53. Jaschke, A. 2001. Artificial ribozymes and deoxyribozymes. *Curr. Opin. Struct. Biol.* 11: 321.
54. Schubert, S., and Kurreck, J. 2004. Ribozyme- and deoxyribozyme-strategies for medi-cal applications. *Curr. Drug Targets* 5: 667.
55. Haner, R., and Hall, J. 1997. The sequence-specific cleavage of RNA by artificial chem-ical ribonucleases. *Antisense Nucleic Acid Drug Dev.* 7: 423.

56. Vlassov, V. V., and Vlassov, A. V. 2004. Cleavage of RNA by imidazole. In *Artificial Nucleases* (Series: Nucleic Acids and Molecular Biology), edited by M. A. Zenkova. Heidelberg: Springer-Verlag, 49.

57. Niittymaki, T., and Lonnberg, H. 2006. Artificial ribonucleases. *Org. Biomol. Chem.* 4: 15.

58. Koroleva, L. S., Serpokrylova, I. Y., Vlassov, V. V., and Silnikov, V. N. 2007. Design and synthesis of metal-free artificial ribonucleases. *Protein Pept. Lett.* 14: 151.

59. Lafarge-Frayssinet, C., Duc, H. T., Frayssinet, C., Sarasin, A., Anthony, D., Guo, Y., and Trojan, J. 1997. Antisense insulin-like growth factor I transferred into a rat hepatoma cell line inhibits tumorigenesis by modulating major histocompatibility complex I cell surface expression. *Cancer Gene Ther.* 4: 276.

60. Varga, L. V., Toth, S., Novak, I., and Falus, A. 1999. Antisense strategies: functions and applications in immunology. *Immunol. Lett.* 69: 217.

61. Chadwick, D. R., and Lever, A. M. 2000. Antisense RNA sequences targeting the 5⊠ leader packaging signal region of human immunodeficiency virus type-1 inhibits viral replication at post-transcriptional stages of the life cycle. *Gene Ther.* 7: 1362.

62. De Backer, M. D., Raponi, M., and Arndt, G. M. 2002. RNA-mediated gene silencing in non-pathogenic and pathogenic fungi. *Curr. Opin. Microbiol.* 3: 323.

63. De Backer, M. D., and Van Dijck, P. 2003. Progress in functional genomics approaches to antifungal drug target discovery. *Trends Microbiol.* 11: 470.

64. Forsyth, R. A., Haselbeck, R. J., Ohlsen, K. L., Yamamoto, R. T., Xu, H., Trawick, J. D., Wall, D., Wang, L., Brown-Driver, V., Froelich, J. M., Kedar, G. C., King, P., McCarthy, M., Malone, C., Misiner, B., Robbins, D., Tan, Z., Zhu, Z. Y., Carr, G., Mosca, D. A., Zamudio, C., Foulkes, J. G., and Zyskind, J. W. 2002. A genome-wide strategy for the identification of essential genes in *Staphylococcus aureus*. *Mol. Microbiol.* 43: 1387.

65. Ji, Y., Woodnutt, G., Rosenberg, M., and Burnham, M. K. 2002. Identification of essential genes in *Staphylococcus aureus* using inducible antisense RNA. *Methods Enzymol.* 358: 123.

66. Ji, Y., Yin, D., Fox, B., Holmes, D. J., Payne, D., and Rosenberg, M. 2004. Validation of antibacterial mechanism of action using regulated antisense RNA expression in *Staphylococcus aureus*. *FEMS Microbiol. Lett.* 231: 177.

67. Yin, D., and Ji, Y. 2002. Genomic analysis using conditional phenotypes generated by antisense RNA. *Curr. Opin. Microbiol.* 5: 330.

68. Yin, D., Fox, B., Lonetto, M. L., Etherton, M. R., Payne, D. J., Holmes, D. J., Rosenberg, M., and Ji, Y. 2004. Identification of antimicrobial targets using a comprehensive genomic approach. *Pharmacogenomics* 5: 101.

69. Noonberg, S. B., Scott, G. K., Garovoy, M. R., Benz, C. C., and Hunt, C. A. 1994. In vivo generation of highly abundant sequence-specific oligonucleotides for antisense and triplex gene regulation. *Nucleic Acids Res.* 22: 2830.

70. Samarsky, D., Ferbeyre, G., and Bertrand, E. 2000. Expressing active ribozymes in cells. *Curr. Issues Mol. Biol.* 2: 87.

71. Sullenger, B. A., and Gilboa, E. 2002. Emerging clinical applications of RNA. *Nature* 418: 252.

72. Thomann, U., and Stange-Thomann, N. 2005. Delivery vectors for short interfering RNA, micro-RNA and antisense RNA. *US Patent Application* 20050203047.

73. Heasman, J. 2002. Morpholino oligos: making sense of antisense? *Dev. Biol.* 243: 209.

74. Delihas, N. 1995. Regulation of gene expression by trans-encoded antisense RNAs. *Mol. Microbiol.* 15: 411.

75. Nellen, W., and Sczakiel, G. 1996. In vitro and in vivo action of antisense RNA. *Mol. Biotechnol.* 6: 7.

76. Knee, R., and Murphy, P. R. 1997. Regulation of gene expression by natural antisense RNA transcripts. *Neurochem. Int.* 31: 379.
77. Terryn, N., and Rouze, P. 2000. The sense of naturally transcribed antisense RNAs in plants. *Trends Plant Sci.* 5: 394.
78. Childs, J. L., Poole, A. W., and Turner, D. H. 2003. Inhibition of *Escherichia coli* RNase P by oligonucleotide directed misfolding of RNA. *RNA* 9: 1437.
79. Disney, M. D., Childs, J. L., and Turner, D. H. 2004. New approaches to targeting RNA with oligonucleotides: inhibition of group I intron self-splicing. *Biopolymers* 73: 151.
80. Enerly, E., Sheng, Z. and Li, K. B. 2005. Natural antisense as potential regulator of alternative initiation, splicing and termination. *In Silico Biol.* 5: 33.
81. Li, M., and Rossi, J. J. 2005. Lentiviral vector delivery of siRNA and shRNA encoding genes into cultured and primary hematopoietic cells. *Methods Mol. Biol.* 309: 261.
82. Sano, M., Kato, Y., and Akashi, H. 2005. Novel methods for expressing RNA interference in human cells. *Methods Enzymol.* 392: 97.
83. Zhang, Y. C., Taylor, M. M., Samson, W. K., and Phillips, M. I. 2005. Antisense inhibition: oligonucleotides, ribozymes, and siRNAs. *Methods Mol. Med.* 106: 11.
84. Di Serio, F., Schob, H., Iglesias, A., Tarina, C., Bouldoires, E., and Meins, F., Jr. 2001. Sense- and antisense-mediated gene silencing in tobacco is inhibited by the same viral suppressors and is associated with accumulation of small RNAs. *Proc. Natl. Acad. Sci. USA* 98: 6506.
85. Martinez, J., Patkaniowska, A., Urlaub, H., Luhrmann, R., and Tuschl, T. 2002. Single-stranded antisense siRNAs guide target RNA cleavage in RNAi. *Cell* 110: 563.
86. Tijsterman, M., Ketting, R. F., Okihara, K. L., Sijen, T., and Plasterk, R. H. 2002. RNA helicase MUT-14-dependent gene silencing triggered in *C. elegans* by short antisense RNAs. *Science* 295: 694.
87. Holen, T., Amarzguioui, M., Babaie, E., and Prydz, H. 2003. Similar behavior of single-strand and double-strand siRNAs suggests they act through a common RNAi pathway. *Nucleic Acids Res.* 31: 2401.
88. Paddison, P. J., and Hannon, G. J. 2003. siRNAs and shRNAs: skeleton keys to the human genome. RNA interference: the new somatic cell genetics? *Curr. Opin. Mol. Ther.* 5: 217.
89. Shi, Y. 2003. Mammalian RNAi for the masses. *Trends Genet.* 19: 9.
90. Bohula, E. A, Salisbury, A. J., Sohail, M., Playford, M. P., Riedemann, J., Southern, E. M., and Macaulay, V. M. 2003. The efficacy of small interfering RNAs targeted to the type 1 IGF receptor is influenced by secondary structure in the IGF1R transcript. *J. Biol. Chem.* 278: 15991.
91. Brown, K. M., Chu, C. Y., and Rana, T. M. 2005. Target accessibility dictates the potency of human RISC. *Nat. Struct. Mol. Biol.* 12: 469.
92. Overhoff, M., Alken, M., Far, R. K., Lemaitre, M., Lebleu, B., Sczakiel, G., and Robbins, I. 2005. Local RNA target structure influences siRNA efficacy: a systematic global analysis. *J. Mol. Biol.* 348: 871.
93. Schubert, S., Grunweller, A., Erdmann, V. A., and Kurreck, J. 2005. Local RNA target structure influences siRNA efficacy: systematic analysis of intentionally designed binding regions. *J. Mol. Biol.* 348: 883.
94. Kitabwalla, M., and Ruprecht, R. M. 2002. RNA interference: a new weapon against HIV and beyond. *New Engl. J. Med.* 347: 1364.
95. Boden, D., Pusch, O., Lee, F., Tucker, L., and Ramratnam, B. 2003. Human immunodeficiency virus type 1 escape from RNA interference. *J. Virol.* 77: 11531.
96. Goldbach, R., Bucher, E., and Prins, M. 2003. Resistance mechanisms to plant viruses: an overview. *Virus Res.* 92: 207.
97. Kubota, K., Tsuda, S., Tamai, A., and Meshi, T. 2003. Tomato mosaic virus replication protein suppresses virus-targeted post-transcriptional gene silencing. *J. Virol.* 77: 11016.

98. Lichner, Z., Silhavy, D., and Burgyan, J. 2003. Double-stranded RNA-binding proteins could suppress RNA interference-mediated antiviral defenses. *J. Gen. Virol.* 84 (Pt. 4): 975.

99. Reed, J. C., Kasschau, K. D., Prokhnevsky, A. I., Gopinath, K., Pogue, G. P., Carrington, J. C., and Dolja, V. V. 2003. Suppressor of RNA silencing encoded by beet yellows virus. *Virology* 306: 203.

100. Qu, F., Ren, T., and Morris, T. J. 2003. The coat protein of turnip crinkle virus suppresses post-transcriptional gene silencing at an early initiation step. *J. Virol.* 77: 511.

101. Vargason, J. M., Szittya, G., Burgyan, J., and Tanaka Hall, T. M. 2003. Size selective recognition of siRNA by an RNA silencing suppressor. *Cell* 115: 799.

102. Das, A. T., Brummelkamp, T. R., Westerhout, E. M., Vink, M., Madiredjo, M., Bernards, R., and Berkhout, B. 2004. Human immunodeficiency virus type 1 escapes from RNA interference-mediated inhibition. *J. Virol.* 78: 2601.

103. Voinnet, O. 2005. Induction and suppression of RNA silencing: insights from viral infections. *Nat. Rev. Genet.* 6: 206.

104. Li, W. X., and Ding, S. W. 2006. Virus counterdefense: diverse strategies for evading the RNA-silencing immunity. *Annu. Rev. Microbiol.* 60: 503.

105. Schutz, S., and Sarnow, P. 2006. Interaction of viruses with the mammalian RNA interference pathway. *Virology* 344: 151.

106. Saxena, S., Jonsson, Z. O., and Dutta, A. 2003. Small RNAs with imperfect match to endogenous mRNA repress translation: implications for off-target activity of siRNA in mammalian cells. *J. Biol. Chem.* 278: 44312.

107. Jackson, A. L., and Linsley, P. S. 2004. Noise amidst the silence: off-target effects of siRNAs? *Trends Genet.* 20: 521.

108. Persengiev, S. P., Zhu, X., and Green, M. R. 2004. Nonspecific, concentration-dependent stimulation and repression of mammalian gene expression by small interfering RNAs (siRNAs). *RNA* 10: 12.

109. Scacheri, P. C., Rozenblatt-Rosen, O., Caplen, N. J., Wolfsberg, T. G., Umayam, L., Lee, J. C., Hughes, C. M., Shanmugam, K. S., Bhattacharjee, A., Meyerson, M., and Collins, F. S. 2004. Short interfering RNAs can induce unexpected and divergent changes in the levels of untargeted proteins in mammalian cells. *Proc. Natl. Acad. Sci. USA* 101: 1892.

110. Snove, O., Jr., and Holen, T. 2004. Many commonly used siRNAs risk off-target activity. *Biochem. Biophys. Res. Commun.* 319: 256.

111. Du, Q., Thonberg, H., Wang, J., Wahlestedt, C., and Liang, Z. 2005. A systematic analysis of the silencing effects of an active siRNA at all single-nucleotide mismatched target sites. *Nucleic Acids Res.* 33: 1671.

112. Lin, X., Ruan, X., Anderson, M. G., McDowell, J. A., Kroeger, P. E., Fesik, S. W., and Shen, Y. 2005. siRNA-mediated off-target gene silencing triggered by a 7 nt complementation. *Nucleic Acids Res.* 33: 4527.

113. Smart, N., Scambler, P. J., and Riley, P. R. 2005. A rapid and sensitive assay for quantification of siRNA efficiency and specificity. *Biol. Proceed. Online* 7: 1.

114. Fedorov, Y., Anderson, E. M., Birmingham, A., Reynolds, A., Karpilow, J., Robinson, K., Leake, D., Marshall, W. S., and Khvorova, A. 2006. Off-target effects by siRNA can induce toxic phenotype. *RNA* 12: 1188.

115. Sledz, C. A., Holko, M., de Veer M. J., Silverman, R. H., and Williams, B. R. 2003. Activation of the interferon system by short-interfering RNAs. *Nat. Cell. Biol.* 5: 834.

116. Kim, D. H., Longo, M., Han, Y., Lundberg, P., Cantin, E., and Rossi, J. J. 2004. Interferon induction by siRNAs and ssRNAs synthesized by phage polymerase. *Nat. Biotechnol.* 22: 321.

117. Pebernard, S., and Iggo, R. D. 2004. Determinants of interferon-stimulated gene induction by RNAi vectors. *Differentiation* 72: 103.

118. Samuel, C. E. 2004. Knockdown by RNAi-proceed with caution. *Nat. Biotechnol.* 22: 280.

119. Judge, A. D., Sood, V., Shaw, J. R., Fang, D., McClintock, K., and Maclachlan, I. 2005. Sequence-dependent stimulation of the mammalian innate immune response by synthetic siRNA. *Nat. Biotechnol.* 23: 457.

120. De Veer, M. J., Sledz, C. A., and Williams, B. R. 2005. Detection of foreign RNA: implications for RNAi. *Immunol. Cell. Biol.* 83: 224.

121. Grimm, D., Streetz, K. L., Jopling, C. L., Storm, T. A., Pandey, K., Davis, C. R., Marion, P., Salazar, F., and Kay, M. A. 2006. Fatality in mice due to oversaturation of cellular microRNA/short hairpin RNA pathways. *Nature* 441: 537.

122. Achenbach, T. V., Brunner, B., and Heermeier, K. 2003. Oligonucleotide-based knockdown technologies: antisense versus RNA interference. *Chembiochem.* 4: 928.

123. Lavery, K. S., and King, T. H. 2003. Antisense and RNAi: powerful tools in drug target discovery and validation. *Curr. Opin. Drug Discov. Dev.* 6: 561.

124. Kramer, R., and Cohen, D. 2004. Functional genomics to new drug targets. *Nat. Rev. Drug Discov.* 3: 965.

125. Scanlon, K. J. 2004. Anti-genes: siRNA, ribozymes and antisense. *Curr. Pharm. Biotechnol.* 5: 415.

126. Corey, D. R. 2007. RNA learns from antisense. *Nat. Chem. Biol.* 3: 8.

127. Iversen, P. L. 2001. Phosphorodiamidate morpholino oligomers: favorable properties for sequence-specific gene inactivation. *Curr. Opin. Mol. Ther.* 3: 235.

128. Summerton, J. E. 2007. Morpholino, siRNA, and S-DNA compared: impact of structure and mechanism of action on off-target effects and sequence specificity. *Curr. Top. Med. Chem.* 7: 651.

129. Al-Anouti, F., and Ananvoranich, S. 2002. Comparative analysis of antisense RNA, double-stranded RNA, and delta ribozyme-mediated gene regulation in *Toxoplasma gondii*. *Antisense Nucleic Acid Drug Dev.* 12: 275.

130. Brantl, S. 2002. Antisense-RNA regulation and RNA interference. *Biochim. Biophys. Acta* 1575: 15.

131. Grunweller, A., Wyszko, E., Bieber, B., Jahnel, R., Erdmann, V. A., and Kurreck, J. 2003. Comparison of different antisense strategies in mammalian cells using locked nucleic acids, 2'-O-methyl RNA, phosphorothioates and small interfering RNA. *Nucleic Acids Res.* 31: 3185.

132. Miyagishi, M., Hayashi, M., and Taira, K. 2003. Comparison of the suppressive effects of antisense oligonucleotides and siRNAs directed against the same targets in mammalian cells. *Antisense Nucleic Acid Drug Dev.* 13: 1.

133. Vickers, T. A., Koo, S., Bennett, C. F., Crooke, S. T., Dean, N. M., and Baker, B. F. 2003. Efficient reduction of target RNAs by small interfering RNA and RNase H-dependent antisense agents. A comparative analysis. *J. Biol. Chem.* 278: 7108.

134. Jepsen, J. S., and Wengel, J. 2004. LNA-antisense rivals siRNA for gene silencing. *Curr. Opin. Drug Discov. Devel.* 7: 188.

135. Gautherot, I., and Sodoyer, R. 2004. A multi-model approach to nucleic acid-based drug development. *BioDrugs* 18: 37.

136. Volloch, V., Schweitzer, B., and Rits, S. 1991. Inhibition of pre-mRNA splicing by antisense RNA in vitro: effect of RNA containing sequences complementary to exons. *Biochem. Biophys. Res. Commun.* 179: 1593.

137. Taylor, J. K., Zhang, Q. Q., Wyatt, J. R., and Dean, N. M. 1999. Induction of endogenous Bcl-xS through the control of Bcl-x pre-mRNA splicing by antisense oligonucleotides. *Nat. Biotechnol.* 17: 1097.

138. Zeng, Y., Gu, X., Chen, Y., Gong, L., Ren, Z., and Huang, S. 1999. Reversal of aberrant splicing of beta-thalassaemia allele by antisense RNA in vitro and in vivo. *Chin. Med. J. (Engl.)* 112: 107.

139. Gong, L., Gu, X. F., Chen, Y. D., Ren, Z. R., Huang, S. Z., and Zeng, Y. T. 2000. Reversal of aberrant splicing of beta-thalassaemia allele (IVS-2-654 C→T) by antisense RNA expression vector in cultured human erythroid cells. *Br. J. Haematol.* 111: 351.

140. Sierakowska, H., Agrawal, S., and Kole, R. 2000. Antisense oligonucleotides as modulators of pre-mRNA splicing. *Methods Mol. Biol.* 133: 223.

141. Mercatante, D. R., Bortner, C. D., Cidlowski, J. A., and Kole, R. 2001. Modification of alternative splicing of Bcl-x pre-mRNA in prostate and breast cancer cells. *J. Biol. Chem.* 276: 16411.

142. Skordis, L. A., Dunckley, M. G., Yue, B., Eperon, I. C., and Muntoni, F. 2003. Bifunctional antisense oligonucleotides provide a trans-acting splicing enhancer that stimulates SMN2 gene expression in patient fibroblasts. *Proc. Natl. Acad. Sci. USA* 100: 4114.

143. Vacek, M., Sazani, P., and Kole. R. 2003. Antisense-mediated redirection of mRNA splicing. *Cell Mol. Life. Sci.* 60: 825.

144. Politz, J. C., Browne, E. S., Wolf, D. E., and Pederson, T. 1998. Intranuclear diffusion and hybridization state of oligonucleotides measured by fluorescence correlation spectroscopy in living cells. *Proc. Natl. Acad. Sci. USA* 95: 6043.

145. Tavitian, B. 2000. In vivo antisense imaging. *Q. J. Nucl. Med.* 44: 236.

146. Molenaar, C., Marras, S. A., Slats, J. C., Truffert, J. C., Lemaitre, M., Raap, A. K., Dirks, R. W., and Tanke, H. J. 2001. Linear 2′-O-Methyl RNA probes for the visualization of RNA in living cells. *Nucleic Acids Res.* 29: E89.

147. Pederson, T. 2001. Fluorescent RNA cytochemistry: tracking gene transcripts in living cells. *Nucleic Acids Res.* 29: 1013.

148. Dirks, R. W., Molenaar, C., and Tanke, H. J. 2003. Visualizing RNA molecules inside the nucleus of living cells. *Methods* 29: 51.

149. Hutvagner, G., Simard, M. J., Mello, C. C., and Zamore, P. D. 2004. Sequence-specific inhibition of small RNA function. *PLoS Biol.* 2: E98.

150. Meister, G., Landthaler, M., Dorsett, Y., and Tuschl, T. 2004. Sequence-specific inhibition of microRNA- and siRNA-induced RNA silencing. *RNA* 10: 544.

151. Krutzfeldt, J., Rajewsky, N., Braich, R., et al. 2005. Silencing of microRNAs in vivo with "antagomirs." *Nature* 438: 685.

152. Davis, S., Lollo, B., Freier, S., and Esau, C. 2006. Improved targeting of miRNA with antisense oligonucleotides. *Nucleic Acids Res.* 34: 2294.

153. Hammond, S. M. 2006. MicroRNA therapeutics: a new niche for antisense nucleic acids. *Trends Mol. Med.* 12: 99.

154. Orom, U. A., Kauppinen, S., and Lund, A. H. 2006. LNA-modified oligonucleotides mediate specific inhibition of microRNA function. *Gene* 372: 137.

155. Weiler, J., Hunziker, J., and Hall, J. 2006. Anti-miRNA oligonucleotides (AMOs): ammunition to target miRNAs implicated in human disease? *Gene Ther.* 13: 496.

156. Hammond, S. M. 2006. MicroRNAs as oncogenes. *Curr. Opin. Genet. Dev.* 16: 4.

157. Pfeffer, S., and Voinnet, O. 2006. Viruses, microRNAs and cancer. *Oncogene* 25: 6211.

158. Wu, W., Sun, M., Zou, G. M., and Chen, J. 2007. MicroRNA and cancer: current status and prospective. *Int. J. Cancer* 120: 953.

159. Cullen, B. R. 2006. Viruses and microRNAs. *Nat. Genet.* 38 (Suppl.): S25.

160. Dykxhoorn, D. M. 2007. MicroRNAs in viral replication and pathogenesis. *DNA Cell Biol.* 26: 239.

161. Pan, X., Zhang, B., San Francisco. M., and Cobb, G. P. 2007. Characterizing viral microRNAs and its application on identifying new microRNAs in viruses. *J. Cell Physiol.* 211: 10.

162. Herschlag, D. 1991. Implications of ribozyme kinetics for targeting the cleavage of specific RNA molecules in vivo: more isn't always better. *Proc. Natl. Acad. Sci. USA* 88: 6921.

163. Monia, B. P., Johnston, J. F., Ecker, D. J., Zounes, M. A., Lima, W. F., and Freier, S. M. 1992. Selective inhibition of mutant Ha-ras mRNA expression by antisense oligonucleotides. *J. Biol. Chem.* 267: 19954.

164. Hougaard, D. M., Hansen, H., and Larson, L.-I. 1997. Non-radioactive in situ hybridization for mRNA with emphasis on the use of oligodeoxynucleotide probes. *Histochem. Cell. Biol.* 108: 335.

165. Majlessi, M., Nelson, N. C., and Becker, M. M. 1998. Advantages of 2'-O-methyl oligoribonucleotide probes for detecting RNA targets. *Nucleic Acids Res.* 26: 2224.

166. Saha, S., Sparks, A. B., Rago, C., Akmaev, V., Wang, C. J., Vogelstein, B., Kinzler, K. W., and Velculescu, V. E. 2002. Using the transcriptome to annotate the genome. *Nat. Biotechnol.* 20: 508.

167. International Human Genome Sequencing Consortium. 2004. Finishing the euchromatic sequence of the human genome. *Nature* 431: 931.

168. Nielsen, P. E. 2000. Peptide nucleic acids: on the road to new gene therapeutic drugs. *Pharmacol. Toxicol.* 86: 3.

169. Bergeron, L. J., Ouellet, J., and Perreault, J. P. 2003. Ribozyme-based gene-inactivation systems require a fine comprehension of their substrate specificities: the case of delta ribozyme. *Curr. Med. Chem.* 10: 2589.

170. Fakler, B., Herlitze, S., Amthor, B., Zenner, H. P., and Ruppersberg, J. P. 1994. Short antisense oligonucleotide-mediated inhibition is strongly dependent on oligo length and concentration but almost independent of location of the target sequence. *J. Biol. Chem.* 269: 16187.

171. Johansson, H. E., Belsham, G. J., Sproat, B. S., and Hentze, M. W. 1994. Target-specific arrest of mRNA translation by antisense 2'-O-alkyloligoribonucleotides. *Nucleic Acids Res.* 22: 4591.

172. Flanagan, W. M., Kothavale, A., and Wagner, R. W. 1996. Effects of oligonucleotide length, mismatches and mRNA levels on C-5 propyne-modified antisense potency. *Nucleic Acids Res.* 24: 2936.

173. Wagner, R. W., Matteucci, M. D., Grant, D., Huang, T., and Froehler, B. C. 1996. Potent and selective inhibition of gene expression by an antisense heptanucleotide. *Nat Biotechnol.* 14: 840.

174. Woolf, T. 1996. It's not the size, it's the potency. *Nat Biotechnol.* 14: 824.

175. Deere, J., Iversen, P., and Geller, B. L. 2005. Antisense phosphorodiamidate morpholino oligomer length and target position effects on gene-specific inhibition in *Escherichia coli. Antimicrob. Agents Chemother.* 49: 249.

176. Roberts, R. W., and Crothers, D. M. 1992. Stability and properties of double and triple helices: dramatic effects of RNA or DNA backbone composition. *Science* 258: 1463.

177. Lesnik, E. A., and Freier, S. M. 1995. Relative thermodynamic stability of DNA, RNA, and DNA:RNA hybrid duplexes: relationship with base composition and structure. *Biochemistry* 34: 10807.

178. Sugimoto, N., Nakano, S., Katoh, M., Matsumura, A., Nakamuta, H., Ohmichi, T., Yoneyama, M., and Sasaki, M. 1995. Thermodynamic parameters to predict stability of RNA/DNA hybrid duplexes. *Biochemistry* 34: 11211.

179. Wu, P., Nakano, S., and Sugimoto, N. 2002. Temperature dependence of thermodynamic properties for DNA/DNA and RNA/DNA duplex formation. *Eur. J. Biochem.* 269: 2821.

180. Southern, E., Mir, K., and Shchepinov, M. 1999. Molecular interactions on microarrays. *Nat. Genet.* 21: 5.

181. Venturini, F., Braspenning, J., Homann, M., Gissmann, L., and Sczakiel, G. 1999. Kinetic selection of HPV 16 E6/E7-directed antisense nucleic acids: anti-proliferative effects on HPV 16-transformed cells. *Nucleic Acids Res.* 27: 1585.

182. Huang, C. Y., Kasai, M., and Buetow, D. E. 1998. Extremely-rapid RNA detection in dot blots with digoxigenin-labeled RNA probes. *Genet. Anal.* 14: 109.

183. Kronenwett, R., Haas, R., and Sczakiel, G. 1996. Kinetic selectivity of complementary nucleic acids: bcr-abl-directed antisense RNA and ribozymes. *J. Mol. Biol.* 259: 632.

184. Patzel, V., zu Putlitz, J., Wieland, S., Blum, H. E., and Sczakiel, G. 1997. Theoretical and experimental selection parameters for HBV-directed antisense RNA are related to increased RNA-RNA annealing. *Biol. Chem.* 378: 539.

185. Patzel, V., and Sczakiel, G. 2000. In vitro selection supports the view of a kinetic control of antisense RNA-mediated inhibition of gene expression in mammalian cells. *Nucleic Acids Res.* 28: 2462.

186. Walton, S. P., Stephanopoulos, G. N., Yarmush, M. L., and Roth, C. M. 2002. Thermodynamic and kinetic characterization of antisense oligodeoxynucleotide binding to a structured mRNA. *Biophys J.* 82: 366.

187. Matveeva, O. V., Mathews, D. H., Tsodikov, A. D., Shabalina, S. A., Gesteland, R. F., Atkins, J. F., and Freier, S. M. 2003. Thermodynamic criteria for high-hit rate antisense oligonucleotide design. *Nucleic Acids Res.* 31: 4989.

188. Ratilainen, T., Holmen, A., Tuite, E., Nielsen, P. E., and Norden, B. 2000. Thermodynamics of sequence-specific binding of PNA to DNA. *Biochemistry* 39: 7781.

189. Braasch, D. A., Liu, Y., and Corey, D. R. 2002. Antisense inhibition of gene expression in cells by oligonucleotides incorporating locked nucleic acids: effect of mRNA target sequence and chimera design. *Nucleic Acids Res.* 30: 5160.

190. Valoczi, A., Hornyik, C., Varga, N., Burgyan, J., Kauppinen, S., and Havelda, Z. 2004. Sensitive and specific detection of microRNAs by northern blot analysis using LNA-modified oligonucleotide probes. *Nucleic Acids Res.* 32: e175.

191. Fluiter, K., Frieden, M., Vreijling, J., Koch, T., and Baas, F. 2005. Evaluation of LNA-modified DNAzymes targeting a single nucleotide polymorphism in the large subunit of RNA polymerase II. *Oligonucleotides* 15: 246.

192. Maher, L. J., 3rd, and Dolnick, B. J. 1988. Comparative hybrid arrest by tandem antisense oligodeoxyribonucleotides or oligodeoxyribonucleoside methylphosphonates in a cell-free system. *Nucleic Acids Res.* 16: 3341.

193. Kandimalla, E. R., Manning, A., Lathan, C., Byrn, R. A., and Agrawal, S. 1995. Design, biochemical, biophysical and biological properties of cooperative antisense oligonucleotides. *Nucleic Acids Res.* 23: 3578.

194. Tanner, N. K. 1999. Ribozymes: the characteristics and properties of catalytic RNAs. *FEMS Microbiol. Rev.* 23: 257.

195. Blount, K. F., and Uhlenbeck, O. C. 2002. The hammerhead ribozyme. *Biochem. Soc. Trans.* 30 (Pt. 6): 1119.

196. Fedor, M. J. 2000. Structure and function of the hairpin ribozyme. *J. Mol. Biol.* 297: 269.

197. Hertel, K. J., Herschlag, D., and Uhlenbeck, O. C. 1996. Specificity of hammerhead ribozyme cleavage. *EMBO J.* 15: 3751.

198. Phylactou, L. A., Tsipouras, P., and Kilpatrick, M. W. 1998. Hammerhead ribozymes targeted to the FBN1 mRNA can discriminate a single base mismatch between ribozyme and target. *Biochem. Biophys. Res. Commun.* 249: 804.

199. Sioud, M. 1997. Effects of variations in length of hammerhead ribozyme antisense arms upon the cleavage of longer RNA substrates. *Nucleic Acids Res.* 25: 333.

200. Hormes, R., Homann, M., Oelze, I., Marschall, P., Tabler, M., Eckstein, F., and Sczakiel, G. 1997. The subcellular localization and length of hammerhead ribozymes determine efficacy in human cells. *Nucleic Acids Res.* 25: 769.

201. Hormes, R., and Sczakiel, G. 2002. The size of hammerhead ribozymes is related to cleavage kinetics: the role of substrate length. *Biochimie* 84: 897.

202. Werner, M., Rosa, E., Nordstrom, J. L., Goldberg, A. R., and George, S. T. 1998. Short oligonucleotides as external guide sequences for site-specific cleavage of RNA molecules with human RNase P. *RNA* 4: 847.

203. Ma, M., Benimetskaya, L., Lebedeva, I., Dignam, J., Takle, G., and Stein, C. A. 2000. Intracellular mRNA cleavage induced through activation of RNase P by nuclease-resistant external guide sequences. *Nat. Biotechnol.* 18: 58.

204. Vary, C. P. 1987. A homogeneous nucleic acid hybridization assay based on strand displacement. *Nucleic Acids Res.* 15: 6883.

205. Li, Q., Luan, G., Guo, Q., and Liang, J. 2002. A new class of homogeneous nucleic acid probes based on specific displacement hybridization. *Nucleic Acids Res.* 30: e5.

206. Roberts, R. W., and Crothers, D. M. 1991. Specificity and stringency in DNA triplex formation. *Proc. Natl. Acad. Sci. USA* 88: 9397.

207. Hertel, K. J., Stage-Zimmermann, T. K., Ammons, G., and Uhlenbeck, O. C. 1998. Thermodynamic dissection of the substrate-ribozyme interaction in the hammerhead ribozyme. *Biochemistry* 37: 16983.

208. Ohmichi, T., and Kool, E. T. 2000. The virtues of self-binding: high sequence specificity for RNA cleavage by self-processed hammerhead ribozymes. *Nucleic Acids Res.* 28: 776.

209. Tyagi, S., and Kramer, F. R. 1996. Molecular beacons: probes that fluoresce upon hybridization. *Nat. Biotechnol.* 14: 303.

210. Bonnet, G., Tyagi, S., Libchaber, A., and Kramer, F. R. 1999. Thermodynamic basis of the enhanced specificity of structured DNA probes. *Proc. Natl. Acad. Sci. USA* 96: 6171.

211. Guo, Z., Liu, Q., and Smith, L. M. 1997. Enhanced discrimination of single nucleotide polymorphisms by artificial mismatch hybridization. *Nat. Biotechnol.* 15: 331.

212. Delihas, N., Rokita, S. E., and Zheng, P. 1997. Natural antisense RNA/target RNA interactions: possible models for antisense oligonucleotide drug design. *Nat. Biotechnol.* 15: 751.

213. Zeiler, B. N., and Simons, R. W. 1998. Antisense RNA structure and function. In *RNA structure and function*, ed. R. W. Simons and M. Grunberg-Manago. Cold Spring Harbor, N.Y.: Cold Spring Harbor Laboratory Press, p. 437.

214. Brantl, S. 2002. Antisense RNAs in plasmids: control of replication and maintenance. *Plasmid* 48: 165.

215. Kikuchi, K., Umehara, T., Fukuda, K., Kuno, A., Hasegawa, T., and Nishikawa, S. 2005. A hepatitis C virus (HCV) internal ribosome entry site (IRES) domain III-IV-targeted aptamer inhibits translation by binding to an apical loop of domain IIId. *Nucleic Acids Res.* 33: 683.

216. Darfeuille, F., Reigadas, S., Hansen, J. B., Orum, H., Di Primo, C., and Toulme, J. J. 2006. Aptamers targeted to an RNA hairpin show improved specificity compared to that of complementary oligonucleotides. *Biochemistry* 45: 12076.

217. Johnston, B. H., Kazakov, S. A., and Kisich, K. 1998. Antisense and antigene therapeutics with improved binding properties and methods for their use. International Patent Application (PCT) WO99/09045.

218. Johnston, B. H., Kazakov, S. A., and Kisich, K. 2003. Antisense and antigene therapeutics with improved binding properties and methods for their use. Australian Patent AU756301.

219. Seyhan, A. A., Alizadeh, B., Austin, I., Kazakov, S. A., Schweitzer, C., Lundstrom, K., and Johnston, B. H. 2001. Virally-expressed RNA Padlocks inhibit TNFα protein synthesis. *Nucleic Acids Res. Symp. Ser.* 42: 62.

220. Kazakov, S. A., Dallas, A., Kuo, T.-C., and Johnston, B. H. 2005. Method of preparation of RNA capable of target-dependent circularization and topological linkage. International Patent Application (PCT) WO2005/001063.

221. Dallas, A., Balatskaya, S. V., Kuo, T.-C., Vlassov, A. V., Kaspar, R. L., Kisich, K. O., Kazakov, S. A., and Johnston, B. H. 2007. Hairpin ribozyme-antisense RNA derivatives act as molecular Lassos. (Submitted), 2008.

222. Feldstein, P. A., and Bruening, G. 1993. Catalytically active geometry in the reversible circularization of "mini-monomer" RNAs derived from the complementary strand of tobacco ringspot virus satellite RNA. *Nucleic Acids Res.* 21: 1991.

223. Komatsu, Y., Koizumi, M., Nakamura, H., and Ohtsuka, E. 1993. Loop-size variation to probe hairpin ribozymes. *J. Am. Chem. Soc.* 116: 3692.

224. Komatsu, Y., Kanzaki, I., Koizumi, M., and Ohtsuka, E. 1995. Modification of primary structures of hairpin ribozymes for probing active conformations. *J. Mol. Biol.* 252: 296.

225. Kool, E. T. 1996. Circular oligonucleotides: new concepts in oligonucleotide design. *Annu. Rev. Biophys. Biomol. Struct.* 25: 1.

226. Rowley, P. T., Kosciolek, B. A., and Kool, E. T. 1999. Circular antisense oligonucleotides inhibit growth of chronic myeloid leukemia cells. *Mol. Med.* 5: 693.

227. Nilsson, M., Malmgren, H., Samiotaki, M., Kwiatkowski, M., Chowdhary, B.P., and Landegren, U. 1994. Padlock probes: circularizing oligonucleotides for localized DNA detection. *Science* 265: 2085.

228. Nilsson, M., Antson, D. O., Barbany, G., and Landegren, U. 2001. RNA-templated DNA ligation for transcript analysis. *Nucleic Acids Res.* 29: 578.

229. Myer, S. E., and Day, D. J. 2001. Synthesis and application of circularizable ligation probes. *Biotechniques* 30: 584.

230. Nilsson, M., Barbany, G., Antson, D. O., Gertow, K., and Landegren, U. 2000. Enhanced detection and distinction of RNA by enzymatic probe ligation. *Nat. Biotechnol.* 18: 791.

231. Far, R. K.-K., and Sczakiel, G. 2003. The activity of siRNA in mammalian cells is related to structural target accessibility: a comparison with antisense oligonucleotides. *Nucleic Acids Res.* 31: 4417.

232. Kumar, R., Conklin, D. S., and Mittal, V. 2003. High-throughput selection of effective RNAi probes for gene silencing. *Genome Res.* 13: 2333.

233. Miller, V. M., Xia, H., Marrs, G. L., Gouvion, C. M., Lee, G., Davidson, B. L., and Paulson, H. L. 2003. Allele-specific silencing of dominant disease genes. *Proc. Natl. Acad. Sci. USA* 100: 7195.

234. Kronke, J., Kittler, R., Buchholz, F., Windisch, M. P., Pietschmann, T., Bartenschlager, R., and Frese, M. 2004. Alternative approaches for efficient inhibition of hepatitis C virus RNA replication by small interfering RNAs. *J. Virol.* 78: 3436.

235. Yoshinari, K., Miyagishi, M., and Taira, K. 2004. Effects on RNAi of the tight structure, sequence and position of the targeted region. *Nucleic Acids Res.* 32: 691.

236. Denman, R. B. 1996. Facilitator oligonucleotides increase ribozyme RNA binding to full-length RNA substrates in vitro. *FEBS Lett.* 382: 116.

237. Jankowsky, E., and Schwenzer, B. 1998. Oligonucleotide facilitators enable a hammerhead ribozyme to cleave long RNA substrates with multiple-turnover activity. *Eur. J. Biochem.* 254: 129.

238. Horn, S., and Schwenzer, B. 1999. Oligonucleotide facilitators enhance the catalytic activity of RNA-cleaving DNA enzymes. *Antisense Nucleic Acid Drug. Dev.* 9: 465.

239. Hovig, E., Maelandsmo, G., Mellingsaeter, T., Fodstad, O., Mielewczyk, S. S., Wolfe, J., and Goodchild, J. 2001. Optimization of hammerhead ribozymes for the cleavage of S100A4 (CAPL) mRNA. *Antisense Nucleic Acid Drug Dev.* 11: 67.

240. Rittner, K., Burmester, C., and Sczakiel, G. 1993. In vitro selection of fast-hybridizing and effective antisense RNAs directed against the human immunodeficiency virus type 1. *Nucleic Acids Res.* 21: 1381.

241. Milner, N., Mir, K. U., and Southern, E. M. 1997. Selecting effective antisense reagents on combinatorial oligonucleotide arrays. *Nat. Biotechnol.* 15: 537.

242. Sczakiel, G. 1997. The design of antisense RNA. *Antisense Nucleic Acid Drug Dev.* 7: 439.

243. Matveeva, O., Felden, B., Audlin, S., Gesteland, R. F., and Atkins, J. F. 1997. A rapid in vitro method for obtaining RNA accessibility patterns for complementary DNA probes: correlation with an intracellular pattern and known RNA structures. *Nucleic Acids Res.* 25: 5010.

244. Ho, S. P., Britton, D. H., Stone, B. A., Behrens, D. L., Leffet, L. M., Hobbs, F. W., Miller, J. A., and Trainor, G. L. 1996. Potent antisense oligonucleotides to the human multidrug resistance-1 mRNA are rationally selected by mapping RNA-accessible sites with oligonucleotide libraries. *Nucleic Acids Res.* 24: 1901.

245. Lloyd, B. H., Giles, R. V., Spiller, D. G., Grzybowski, J., Tidd, D. M., and Sibson, D. R. 2001. Determination of optimal sites of antisense oligonucleotide cleavage within TNF-alpha mRNA. *Nucleic Acids Res.* 29: 3664.

246. Sohail, M., and Southern, E. M. 2000. Selecting optimal antisense reagents. *Adv. Drug Deliv. Rev.* 44: 23.

247. Wrzesinski, J., Legiewicz, M., and Ciesiolka, J. 2000. Mapping of accessible sites for oligonucleotide hybridization on hepatitis delta virus ribozymes. *Nucleic Acids Res.* 28: 1785.

248. Allawi, H. T., Dong, F., Ip, H. S., Neri, B. P., and Lyamichev, V. I. 2001. Mapping of RNA accessible sites by extension of random oligonucleotide libraries with reverse transcriptase. *RNA* 7: 314.

249. Scherr, M., LeBon, J., Castanotto, D., Cunliffe, H. E., Meltzer, P. S., Ganser, A., Riggs, A. D., and Rossi, J. J. 2001. Detection of antisense and ribozyme accessible sites on native mRNAs: application to NCOA3 mRNA. *Mol. Ther.* 4: 454.

250. Sohail, M., Hochegger, H., Klotzbucher, A., Guellec, R. L., Hunt, T., and Southern, E. M. 2001. Antisense oligonucleotides selected by hybridisation to scanning arrays are effective reagents in vivo. *Nucleic Acids Res.* 29: 2041.

251. Giddings, M. C., Shah, A. A., Freier, S., Atkins, J. F., Gesteland, R. F., and Matveeva, O. V. 2002. Artificial neural network prediction of antisense oligodeoxynucleotide activity. *Nucleic Acids Res.* 30: 4295.

252. Gabler, A., Krebs, S., Seichter, D., and Forster, M. 2003. Fast and accurate determination of sites along the FUT2 in vitro transcript that are accessible to antisense oligonucleotides by application of secondary structure predictions and RNase H in combination with MALDI-TOF mass spectrometry. *Nucleic Acids Res.* 31: e79.

253. Pan, W. H., and Clawson, G. A. 2006. Identifying accessible sites in RNA: the first step in designing antisense reagents. *Curr. Med. Chem.* 13: 3083.

254. Sun, Y., Duan, M., Lin, R., Wang, D., Li, C., Bo, X., and Wang, S. 2006. A novel integrated strategy (full length gene targeting) for mRNA accessible site tagging combined with microarray hybridization/RNase H cleavage to screen effective antisense oligonucleotides. *Mol. Vis.* 12: 1364.

255. Crisell, P., Thompson, S., and James, W. 1993. Inhibition of HIV-1 replication by ribozymes that show poor activity in vitro. *Nucleic Acids Res.* 21: 5251.

256. Laptev, A. V., Lu, Z., Colige, A., and Prockop, D. J. 1994. Specific inhibition of expression of a human collagen gene (COL1A1) with modified antisense oligonucleotides. The most effective target sites are clustered in double-stranded regions of the predicted secondary structure for the mRNA. *Biochemistry* 33: 11033.

257. Yu, Q., Pecchia, D. B., Kingsley, S. L., Heckman, J. E., and Burke, J. M. 1998. Cleavage of highly structured viral RNA molecules by combinatorial libraries of hairpin ribozymes. The most effective ribozymes are not predicted by substrate selection rules. *J. Biol. Chem.* 273: 23524.

258. Scherr, M., Rossi, J. J., Sczakiel, G., and Patzel, V. 2000. RNA accessibility prediction: a theoretical approach is consistent with experimental studies in cell extracts. *Nucleic Acids Res.* 28: 2455.

259. Sczakiel, G., and Far, R. K. 2002. The role of target accessibility for antisense inhibition. *Curr. Opin. Mol. Ther.* 4: 149.

260. Lieber, A., and Strauss, M. 1995. Selection of efficient cleavage sites in target RNAs by using a ribozyme expression library. *Mol. Cell. Biol.* 15: 540.

261. Kramer, F. R., Dubnau, D., Drilka, K. A., and Pinter, A. 1997. Selection of ribozymes that efficiently cleave target RNA. United States Patent US5616459.

262. Kruger, M., Beger, C., Li, Q. X., Welch, P. J., Tritz, R., Leavitt, M., Barber, J. R., and Wong-Staal, F. 2000. Identification of eIF2Bgamma and eIF2gamma as cofactors of hepatitis C virus internal ribosome entry site-mediated translation using a functional genomics approach. *Proc. Natl. Acad. Sci. USA* 97: 8566.

263. Ho, S. P., Bao, Y., Lesher, T., Malhotra, R., Ma, L. Y., Fluharty, S. J., and Sakai, R. R. 1998. Mapping of RNA accessible sites for antisense experiments with oligonucleotide libraries. *Nat. Biotechnol.* 16: 59.

264. Pierce, M. L., and Ruffner, D. E. 1998. Construction of a directed hammerhead ribozyme library: towards the identification of optimal target sites for antisense-mediated gene inhibition. *Nucleic Acids Res.* 26: 5093.

265. Ruffner, D. E., Pierce, M. L., and Chen, Z. 1999. Directed antisense libraries. International Application (PCT) WO99/50457.

266. Sen, G., Wehrman, T. S., Myers, J. W., and Blau, H. M. 2004. Restriction enzyme-generated siRNA (REGS) vectors and libraries. *Nat. Genet.* 36: 183.

267. Shirane, D., Sugao, K., Namiki, S., Tanabe, M., Iino, M., and Hirose, K. 2004. Enzymatic production of RNAi libraries from cDNAs. *Nat. Genet.* 36: 190.

268. Vlassov, A. V., Koval, O. A., Johnston, B. J., and Kazakov, S. A. 2004. ROLL: A method of preparation of gene-specific oligonucleotide libraries. *Oligonucleotides* 14: 210.

269. Seyhan, A. A., Vlassov, A. V., Ilves, H., Egry, L., Kaspar, R. L., Kazakov, S. A., and Johnston, B. H. 2005. Complete, gene-specific siRNA libraries: production and expression in mammalian cells. *RNA* 11: 837.

270. Kazakov, S. A., Vlassov, A. V., Dallas, A., Seyhan, A. A., Egry, L. A., Ilves, H., Kaspar, R. L., and Johnston, B. H. 2006. Methods of preparation of gene-specific oligonucleotide libraries and uses thereof. International Patent Application (PCT) WO2006/007569.

271. Barber, J., Welch, P., Yei, S., and Tritz, R. 1998. Gene functional analysis and discovery using randomized or target-specific ribozyme gene vector libraries. International Patent Application (PCT) WO9832880.

272. Beger, C., Pierce, L. N., Kruger, M., Marcusson, E. G., Robbins, J. M., Welcsh, P., Welch, P. J., Welte, K., King, M. C., Barber, J. R., and Wong-Staal, F. 2001. Identification of Id4 as a regulator of BRCA1 expression by using a ribozyme-library-based inverse genomics approach. *Proc. Natl. Acad. Sci. USA* 98: 130.

273. Miyagishi, M., Matsumoto, S., Akashi, H., Kawasaki, H., Fukao, T., Fukuda, Y., Sano, M., Kato, Y., Takagi, Y., Tanaka, Y., Warashina, M., Kuwabara, T., Sawata, S. Y., Ikeda, Y., Kawahara, S., Sunil, K. C., Wadhwa, R., and Taira, K. 2005. Chemistry-based RNA technologies: demonstration of usefulness of libraries of ribozymes and short hairpin RNAs (shRNAs). *Nucleic Acids Symp. Ser.* 49: 91.

274. Kasim, V., Taira, K., and Miyagishi, M. 2006. Screening of siRNA target sequences by using fragmentized DNA. *J. Gene Med.* 8: 782.

275. Berzal-Herranz, A., and Burke, J. M. 1997. Ligation of RNA molecules by the hairpin ribozyme. *Methods Mol. Biol.* 74: 349.

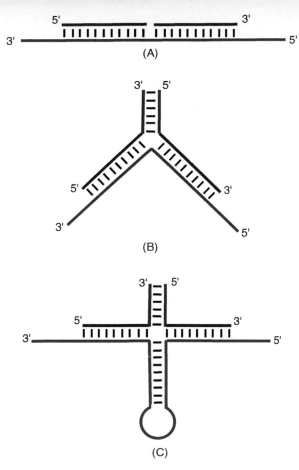

FIGURE 3.1 Cooperative binding of two short oligonucleotides to RNA targets. (A) Side-by-side binding of two oligonucleotides to adjacent target sequences. The complex is stabilized through stacking interactions at the interface between the oligonucleotides. (B) Side-by-side binding of two partially complementary oligonucleotides to adjacent target sequences. The complex is stabilized through base-pairing between the oligonucleotide dimerization segments. (C) Binding of two partially complementary oligonucleotides to nonadjacent target sequences that are brought together in space by a secondary structure in the target. This complex is also stabilized through base-pairing between the oligonucleotide dimerization segments. RNA targets are shown in blue, antisense in red, and the dimerization segments in green.

FI(
(A)
co
co
ma
en
of
wh
en
la
th
to
sta
fo
str

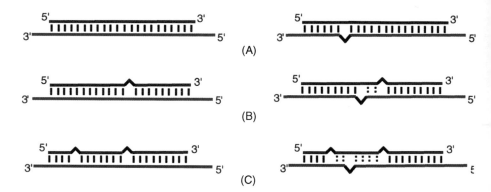

FIGURE 3.4 Perfect and mismatched duplexes between an antisense oligonucleotide and an RNA target. (A) Conventional allele-specific hybridization of a "perfect" antisense with either a normal target (left) or one with a single-nucleotide substitution due to a mutation or SNP (right). (B) Hybridization of a single-base mismatched oligonucleotide with the normal (left) and mutated targets (right). (C) Hybridization of doubly mismatched oligonucleotide with normal (left) and mutated targets (right). In all cases the oligonucleotide forms a more stable duplex with the normal target than the mutated target; however, because mismatches spaced a certain distance apart are especially destabilizing, the presence of two or three mismatches between antisense oligonucleotide and the target may provide better discrimination between the two targets. RNA targets are shown in blue, antisense in red, and mismatches in black. The interactions between complementary bases that are weakened by the nearby mismatches are shown as dotted lines.

Hybridized Linear DNA Padlock

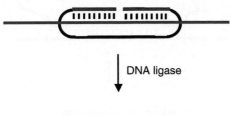

DNA ligase

Hybridized Circular DNA Padlock

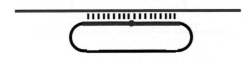

Topologically linked DNA Padlock

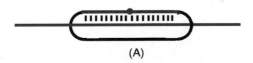

(A)

FIGURE 3.6 Circularizable nucleic acid agents. (A) Padlock Probe (DNA). (B) RNA Lasso. These agents are linear polynucleotides that can hybridize by their antisense segments (shown in red) to an RNA target (blue). Their terminal sequences are joined by either DNA ligase (Padlock Probe) or self-ligated by the encoded ribozyme (RNA Lasso). Note that the ligation site (dot) for the Padlock Probe lies within the antisense-target duplex, whereas for the RNA Lasso it is outside this duplex. Circularization of linear forms of these agents prebound to their targets results in the formation of topologically linked complexes. (Continued on next page.)

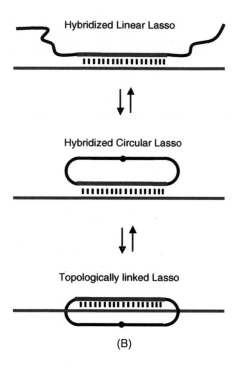

(B)

FIGURE 3.6 (Continued.)

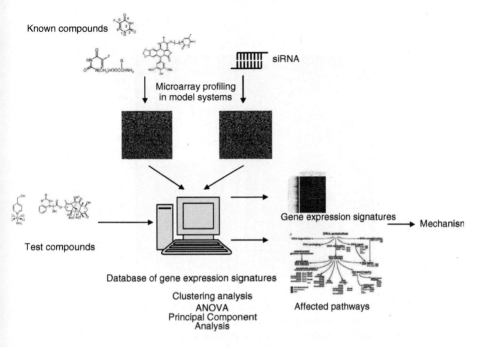

FIGURE 4.2 A genomics database for compound selection and optimization. The database is populated with gene profiles for compounds with known mechanisms of action as well as siRNA. The profiles for novel compounds are then clustered alongside the reference compounds and siRNA to make conclusions about their mechanism. (Reprinted with permission from *Preclinical Development Handbook–Toxicology.* D. Semizarov and E. A. G. Blomme. Shayne Cox Gad, ed. Genomics, 801–839. ©2008. Hoboken, NJ: John Wiley & Sons.)

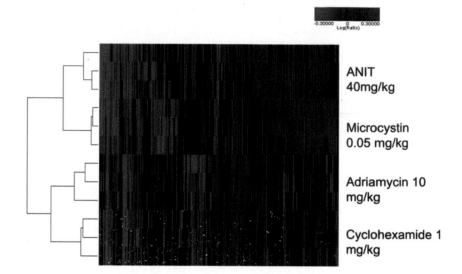

FIGURE 4.3 Hierarchical clustering of gene profiles from the livers of rats treated for 3 days with various hepatotoxicants at toxic doses. Included in the heatmap are genes that were up- or down-regulated by a factor of ≥2 with a *p* value less than 0.01. Green indicates down-regulation, while red indicates up-regulation. Despite significant variability in responses observed with clinical pathology and histopathology, there is limited interindividual variability in gene expression profiles.

4 Gene Profiles in Drug Development

Applications in Target Identification, Biomarker Discovery, and Compound Characterization

Eric A. G. Blomme and Dimitri Semizarov
Global Pharmaceutical Research and Development, Abbott

CONTENTS

4.1 INTRODUCTION

In the past decade, the cost of drug discovery and development has risen exponentially, but the rate of introduction of new drugs has actually decreased. The cost of developing a new chemical entity (NCE) currently ranges from $800 million to $1.1 billion [1, 2]. These new economic fundamentals of the industry are not sustainable, and the problem is being addressed by pharmaceutical companies [2]. Application of novel genomics technologies has offered significant promise in improving the efficiency of the drug discovery and development process. In the years following the publication of the human genome sequence, genomics technologies have revolutionized drug discovery and development. In contrast to the traditional trial-and-error approach, an innovative, hypothesis-driven, and systematic strategy has been adopted. This strategy is based on an initial target selection and validation followed by optimization and selection of compounds that would modulate the activity of the target with minimal side effects. Today's drug development process is characterized by a very high failure rate of experimental compounds, particularly at the costly late stages of development (Phases IIb and III) [3]. In particular, the major causes of attrition in the clinic are currently lack of efficacy and safety, both accounting for approximately 30% of failures [3]. Consequently, approaches using genomics technologies and designed to evaluate or predict toxicity and efficacy earlier represent a solution to control the failure rate in clinical development.

Microarray-based gene profiling currently plays an important role at all stages of drug discovery and development (Figure 4.1). Chronologically, the first area for gene profiling applications is the identification of novel therapeutic targets. Target discovery in all therapeutic areas involves genomic screens to identify disease-causing or disease-modifying genes. For example, identification of therapeutic targets in cancer often starts with a large-scale screen for genes or their products that alter cell proliferation and survival. Identification of genes frequently overexpressed in cancer may reveal a target for therapeutic intervention, whose inhibition by small molecules or antibodies may deprive the cancer cell of its proliferation/survival advantage [4–6]. Later in the discovery pipeline, gene profiling often assists in lead selection and compound optimization. It is now routinely used to profile compounds in order to determine the activated pathways and thus delineate on-target and off-target effects, as well as to characterize toxicological attributes of compounds. In particular, large reference databases are being created to correlate the chemical structures of compounds with their toxicological properties and their gene profiles. As will be illustrated later, such reference databases are critical for the development of predictive expression signatures of toxicity for compound characterization and rank-ordering during the lead selection and lead optimization stages.

Another important application for gene profiling is biomarker discovery. The increasing importance of biomarkers is closely connected with the exponential growth in R&D costs currently experienced by the pharmaceutical industry [7]. An early discovery of a pharmacodynamic biomarker could substantially decrease development costs and cut the development time, thus allowing an earlier market entry and hence an improved patent life cycle [8]. So-called patient stratification biomarkers (i.e., markers correlated with the disease type or response to a drug) are

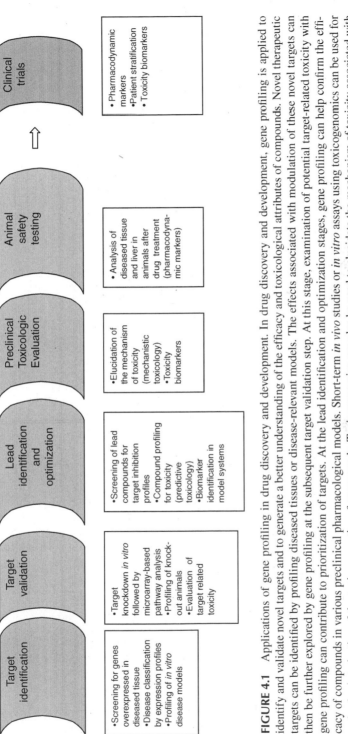

FIGURE 4.1 Applications of gene profiling in drug discovery and development. In drug discovery and development, gene profiling is applied to identify and validate novel targets and to generate a better understanding of the efficacy and toxicological attributes of compounds. Novel therapeutic targets can be identified by profiling diseased tissues or disease-relevant models. The effects associated with modulation of these novel targets can then be further explored by gene profiling at the subsequent target validation step. At this stage, examination of potential target-related toxicity with gene profiling can contribute to prioritization of targets. At the lead identification and optimization stages, gene profiling can help confirm the efficacy of compounds in various preclinical pharmacological models. Short-term in vivo studies or in vitro assays using toxicogenomics can be used for early characterization of the toxic properties of compounds. Toxicogenomics can also be used to elucidate the mechanism of toxicity associated with compounds. These mechanistic studies are typically conducted after candidate selection, but can also be used to establish appropriate counterscreens for backup compound selection. Finally, gene profiling can identify biomarkers for confirming the activity in the clinic (pharmacodynamic markers), stratifying patients in clinical trials, as well as identifying and monitoring toxicological changes, in both preclinical and clinical studies.

particularly valuable in oncology development, as cancer represents a heterogeneous genetic disease. In one of the recent successful examples of patient stratification strategies, the response to Herceptin, an antibody against the *HER-2* gene product, was found to correlate with the amplification of the *HER-2* gene [9–11]. Biomarkers of this type offer the possibility of rationally selecting patients for clinical trials and are now being pursued early in the discovery process.

In addition to monitoring efficacy, biomarkers also offer the opportunity to monitor toxicity in preclinical and clinical studies. In the last few years, toxicologists have used genomics technologies to identify novel biomarkers that would enable early detection of toxicity. These biomarkers are typically in the form of gene expression signatures that can be used *in vitro* and *in vivo* as semiquantitative or quantitative indicators of the development or presence of a toxic change. These toxicological signatures are especially useful in the early stages of discovery to select candidates with optimal toxicological profiles. This improved characterization of compounds at the toxicological level in discovery offers a unique opportunity to reduce the failure rates of compounds at the late stages of development.

In this chapter, we provide an overview of the applications of gene profiling in target identification, compound selection and optimization, early assessment of compound toxicity, and biomarker discovery. A major focus of the chapter is on the impact of gene profiling on the overall efficiency of the drug discovery process and on the success rate of experimental compounds in the late phases of development.

4.2 GENE PROFILING IN TARGET IDENTIFICATION AND VALIDATION

In the past 5–10 years, a growing number of new drugs have been discovered using a target-based approach, illustrating the significant paradigm shift in the pharmaceutical industry [12]. This shift has been facilitated by the decoding of the complete sequence of the human genome, providing access to a large number of potentially interesting therapeutic targets [13–15]. Today, a significant portion of the R&D effort in the pharmaceutical industry is focused on drug target identification and validation, two areas where genomics technologies have already proven their value.

A common strategy for target identification consists of a genome-wide scan for genes or their protein products overexpressed in the diseased tissue relative to its normal counterpart. Gene-expression profiling is an ideal technology for this type of analysis because it provides whole-genome coverage and enables efficient screening of hundreds of samples using a standard protocol. The use of gene profiles is particularly common in the area of cancer target identification.

The development of sophisticated statistical approaches to microarray data analysis, such as supervised clustering, has permitted classification of human cancers on the genomic basis and identification of gene profiles characteristic of subsets of cancers. Hematological cancers were chronologically the first group of cancers to be classified based on genomic profiles, primarily due to better availability of homogeneous tumor cell populations. In a pioneering study of 72 acute leukemia samples, a gene-expression profile was identified that reliably determined the disease type [16]. The profile correctly assigned the samples to the acute myeloid leukemia (AML) or acute lymphoid leukemia (ALL) classes and further defined the subtypes in the ALL category (B-cell and T-cell) [17]. In another study, 360 ALL blasts were analyzed to discover gene-expression profiles associated with the six known clinical subgroups of the disease [18]. While these cancer classification studies produced robust gene-expression-based disease classifiers, no members of these classifiers were pursued as therapeutics targets. A more recent gene-expression study of mixed-lineage leukemia overcame this limitation [19, 20]. Armstrong et al. not only identified a distinct gene-expression pattern that distinguishes MLL from other leukemias, but also identified a single gene that is consistently up-regulated in this leukemia type and explored it as a therapeutic target [19, 20]. The MLL has been historically defined as a subtype based on the presence of specific chromosomal rearrangements, the so-called MLL translocations. The first study established a globally distinct nature of this disease by identifying its characteristic genomic profile [19]. The gene profile of MLL was then analyzed to identify a gene consistently overexpressed in most of the tested MLL samples compared with other acute leukemia samples [20]. This gene encodes FLT3, a receptor tyrosine kinase with an important role in hematopoietic development. An inhibitor of FLT3 was shown to differentially kill cells carrying MLL translocations *in vitro* and inhibit the development of the disease in a mouse model, thus validating FLT3 as a therapeutic target for MLL. This study demonstrates the power of gene-expression profiling in target discovery.

The utility of gene profiles is not limited to hematological cancers. A number of solid tumors have been profiled to identify disease subtypes and discover therapeutic targets. For example, microarray profiling of medulloblastoma, a highly invasive tumor of the cerebellum, led to the identification of expression signatures of the metastatic and nonmetastatic subtypes of the disease [21]. A number of important genes, including platelet-derived growth factor receptor A and several members of the downstream RAS/mitogen-activated protein kinase pathway, were shown to be overexpressed in the metastatic subtype. Thus, these gene profiles identified potential therapeutic targets for the metastatic subtype of medulloblastoma. In another study, over 50 neoplastic and normal prostate samples and 3 common prostate cell lines were profiled using DNA microarrays [22]. Characteristic gene profiles were discovered for localized prostate cancer, metastatic hormone-refractory prostate cancer, and benign prostatic hyperplasia. Two genes consistently showed higher expression levels in the cancer samples, namely, hepsin (a transmembrane serine protease) and PIM-1 (a serine/threonine kinase). The expression of the proteins encoded by these genes was studied using a tissue microarray containing 738 malignant and benign tissues and found to be elevated in prostate cancer. This particular study is a good illustration of the utility of microarray data, since the key findings were confirmed at the protein expression level. A separate gene-profiling study of 11 prostate cancer samples and 4 normal prostate specimens also revealed higher expression levels of hepsin in prostate cancer and thus implicated hepsin as a promising therapeutic target. Later studies have confirmed the validity of hepsin as a cancer target [23, 24].

Although a direct comparison of gene profiles of diseased and normal tissue has resulted in the identification of several therapeutic targets, the utility of gene profiles is not limited to this direct approach. The majority of the genes contained in disease signatures code for proteins that do not belong to the known druggable target classes (i.e., kinases, G protein-coupled receptors, etc.). Furthermore, gene expression changes typically represent the bottom part of the signaling pathways modulating the disease process. Therefore, evaluating regulated pathways in contrast to single genes may represent a more effective approach for target identification. Development of pathway mapping software has facilitated microarray-based target identification by allowing researchers to determine the pathways activated in diseased tissues [25–29]. The pathways highlighted by microarray studies can then be mined for druggable targets. Subsequent target validation steps may involve modulation of the target activity *in vitro* or *in vivo,* followed by microarray profiling to determine whether the target-related signature is affected. Short-interfering RNA is currently the most often used tool for target modulation *in vitro* [30, 31].

All of the studies cited above have been conducted using patient samples. However, clinical samples of sufficient quality are often unavailable or difficult to obtain. Therefore target identification has also been conducted in preclinical model systems, such as tissue culture or animal models. It is important to emphasize that screening of model systems is associated with a high false-positive rate. Indeed, cancer cell lines accumulate a lot of secondary chromosomal aberrations (such as gene amplifications/deletions), which are not relevant to the initial disease-causing events. This necessitates inventive experimental design and careful validation of the results of microarray studies. In an example from the oncology area, a rescue screen was performed in neuroblastoma cells for approximately 900 known therapeutic agents [32]. In this study, 26 drugs were capable of rescuing neuroblastoma cells from oxidant stress. Gene profiles were generated for these therapeutic agents in an attempt to identify a common signature. One of the genes is the common signature coded for a secreted peptide called galanin. In a series of validation experiments, galanin reversed cell death caused by oxidant stress, suggesting that the galanin receptor may represent a therapeutic target. In several other studies, drug targets have been discovered in preclinical models by analyzing pathways activated by the expression of known oncogenes [33, 34].

Although gene profiles can reliably associate gene overexpression with disease, they do not establish a cause-effect relationship between the two. Typically, microarray-based experiments simply generate lists of overexpressed genes and thus necessitate complex follow-up experiments aimed at identifying the genes that play a causative role in the disease process. Therapeutic targets are then validated in model systems using a loss-of-function or gain-of-function approach. Gene profiles can be used at this stage to determine the global effects of target modulation. An example of such an application is a study where the protein kinase *RIalpha gene* was knocked down with an antisense oligonucleotide in cultured cells [35]. Suppression of the protein kinase *RIalpha gene* generated gene expression signatures associated with cell growth, differentiation, and activation pathways. The genes that composed the proliferation-transformation signature were down-regulated, whereas those that defined the differentiation signature were up-regulated in antisense-treated cancer cells and tumors, but not in host livers. These findings validated the proposed mechanism of action of protein kinase RIalpha. Gene silencing by siRNA (small interfering RNA) followed by gene profiling and pathway analysis is a promising tool for target validation [30, 31]. Early studies have demonstrated the feasibility of this approach [36,

37]. We expect that this approach will be actively used in target validation in the near future.

4.3 GENE PROFILING IN COMPOUND CHARACTERIZATION

Genomics technologies, in particular DNA microarrays, have recently become an important tool in compound assessment. Since gene profiles represent a genome-wide view of the physiological state of the cell, they are well suited for characterization of compounds (both in terms of efficacy and toxicity) and ascertaining their mechanism of action. A global approach to the characterization requires the creation of a database of gene profiles for compounds with known mechanisms using a therapeutically relevant model system (Figure 4.2). The database can then be used as a lookup table: after a series of novel compounds is profiled by microarrays, their gene profiles are plugged into the table, and the compounds are clustered based on their gene profiles. One can envision that agents with similar mechanisms will cluster closely, thus enabling the characterization of the test compounds. Several early studies have used this approach to elucidate the mechanism of known and novel agents. For example, three known inhibitors of histone deacetylase (HDAC) were profiled in

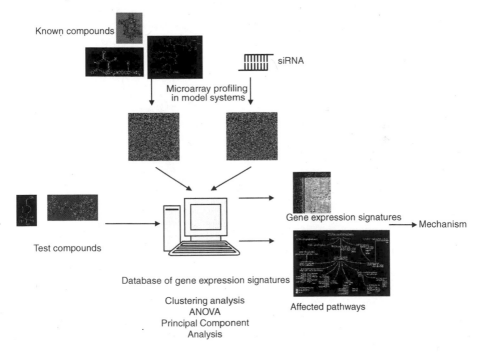

FIGURE 4.2 (See color insert following page 56.) A genomics database for compound selection and optimization. The database is populated with gene profiles for compounds with known mechanisms of action as well as siRNA. The profiles for novel compounds are then clustered alongside the reference compounds and siRNA to make conclusions about their mechanism. (Reprinted with permission from *Preclinical Development Handbook–Toxicology*. D. Semizarov and E. A. G. Blomme. Shayne Cox Gad, ed. Genomics, 801–839. ©2008. Hoboken, NJ: John Wiley & Sons.)

two cell lines to generate a gene profile for HDAC inhibition [38]. The gene profiles of the three potent HDAC inhibitors clustered closely with each other and separately from the profiles for their inactive analogues, implying that the expression signatures are mechanism-based. A core signature of 13 genes was identified that was common to all the HDAC inhibitors in both cell lines. The inhibitors were structurally different, implying that the signatures reflect the HDAC inhibition mechanism rather than off-target effects of the compounds.

Sulindac sulfide, a potent inhibitor of cyclooxygenase 1 (COX-1) and cyclooxygenase 2 (COX-2), has been examined in colorectal carcinoma cells, which express the COX-1 enzyme but very little COX-2 [39]. Microarray profiling identified a group of 11 genes, and their expression was further analyzed in another colon cancer cell line (HCT116) that expresses very low levels of both COX enzymes. Sulindac sulfide had no effect on the expression of the 11-gene "COX signature" in HCT116, implying its association with COX inhibition. Several known COX inhibitors have been shown to induce growth arrest and apoptosis in colon cancer cells. Gene-profiling studies such as the one described above could help elucidate the mechanism of this anticancer effect. This knowledge would enable either selection of one of the known COX inhibitors for preclinical studies or optimization of the structure and creation of new COX inhibitors with potential antitumor activity. More compound-profiling studies are analyzed in a recent comprehensive review [40].

Similar compound-profiling approaches have been used in therapeutic areas other than oncology. In an example from the neuroscience field, gene profiling has been performed for several classes of antidepressants, antipsychotics, and opioid receptor agonists in primary human neurons [41]. The obtained signatures were used to develop statistical models for the prediction of drug efficacy based on gene profiles. The authors demonstrated that several supervised classification algorithms accurately predict the function of each of the drugs.

These studies have demonstrated that gene profiling can answer important questions concerning the drug mechanism and efficacy. The methodology can also be used to investigate drug toxicity if an appropriate model system is used. This aspect is covered more extensively later in this chapter.

4.4   GENE PROFILING IN BIOMARKER DISCOVERY

Identification of efficacy and toxicity biomarkers is a well-established application of gene profiling, which often parallels and facilitates compound selection. A biomarker has been defined as "a characteristic that is objectively measured and evaluated as an indicator of normal biologic processes, pathogenic processes, or pharmacologic responses to a therapeutic intervention" [42]. The increasing importance of biomarkers is closely related to the sharp increase in R&D costs currently experienced by the pharmaceutical industry [7]. An early discovery of a pharmacodynamic biomarker could significantly cut the development time, thus reducing costs and enabling an earlier market entry [8]. Patient-stratification biomarkers (i.e., markers associated with the disease subtype or drug sensitivity) have had the most notable success in oncology development, due to the genetic heterogeneity of cancers. In one of the successful examples of patient-stratification strategies, the response to Herceptin was found to correlate with the amplification of the *HER-2* gene [9–11]. Stratification biomarkers enable the rational selection of patients for clinical trials and are now pursued at the discovery stage. For example, cell culture screens for target inhibition are now accompanied by detailed genomic analysis of the cell lines to identify genetic markers correlated with the response in the assay.

The gene-profiling studies reviewed in the previous subsection may lead to the discovery of mechanism-based pharmacodynamic biomarkers. However, identification of patient-stratification biomarkers typically involves the generation of basal gene expression profiles of the model system, followed by correlation of the profiles with the sensitivity of the system to the drug. Early studies correlating gene profiles with drug sensitivity have been conducted in cultured cell lines [43–45]. Basal gene expression signatures were generated for a panel of 60 cell lines used by the National Cancer Institute for drug discovery screens (NCI-60 panel) [43]. The gene profiles were then correlated with the sensitivity of the cells to several common drugs, such as 5-fluorouracil and L-asparaginase [46]. The basal expression levels of certain genes were then correlated to the mechanisms of drug resistance. Gene profiles were used to predict the chemosensitivity of the cell lines. Staunton et al. designed an algorithm that predicted the sensitivity of the cell lines in the NCI-60 panel to 232 known agents [45]. All 232 compounds were profiled in the 60 cell lines, and the data were divided into a training set and a test set. The training set was used to develop classifiers, and the test set was used to test the accuracy of the classifiers. The study yielded accurate classifiers ($p \leq .05$) and thus proved the feasibility of chemosensitivity prediction by microarrays.

The predictive power of gene profiles for drug sensitivity has also been determined in more complex experimental systems, such as mouse models of cancer [47, 48]. In a recent study, 85 cancer xenografts derived from 9 human organs were profiled by microarrays and examined for sensitivity to 9 anticancer drugs (5-fluorouracil, 3-[(4-amino-2-methyl-5-pyrimidinyl)methyl]-1-(2-chloroethyl)-1-nitrosourea hydrochloride, adriamycin, cyclophosphamide, cisplatin, mitomycin C, methotrexate, vincristine, and vinblastine) [47]. Gene profiles were then generated for all xenografts and correlated with the drug sensitivity. Using the observed correlations, the authors developed an algorithm to calculate the sensitivity score based on the gene expression pattern.

The early studies described above have established the feasibility of predicting drug response to the drug based on the gene profile. However, it remains to be proven that correlations established in model systems will reproduce in patient samples. In contrast to samples from preclinical models, patient samples are more heterogeneous and complex, making the task more challenging. Furthermore, cell culture and xenograft-based models have additional limitations. Most importantly, cultured cells undergo multiple rounds of selection, acquire additional genetic abnormalities, and therefore may poorly represent the original tumor, both in terms of its genomic pattern and its drug sensitivity profile. To date, only a limited number of studies have correlated drug response with the basal expression profile in human samples [49–53]. Additional studies will be required to demonstrate whether gene-expression profiles obtained in preclinical model systems can be used to predict patient response in clinical trials.

4.5 TOXICOGENOMICS: PREDICTING TOXICITY WITH BETTER ACCURACY

4.5.1 DEFINITION OF TOXICOGENOMICS

Microarray technology has rapidly been adopted by pharmaceutical organizations as a useful tool for elucidating toxic mechanisms and identifying compounds with improved toxicological attributes in a faster, more cost-effective manner [54–60]. The term *toxicogenomics* refers to the use of gene-expression profiling in the field of toxicology. Typically, tissues collected from *in vivo* studies or cells derived from *in vitro* experiments are profiled using DNA microarrays. Other technologies for gene profiling are, however, available, and more of them will undoubtedly be developed in the coming years. In particular, for specific applications such as *in vitro* profiling, more cost-effective platforms with a higher throughput are likely to become the preferred tool [61, 62].

Toxicogenomics is based on the finding that toxicants acting through a similar mechanism of action generate similar gene-expression profiles or, at least, affect similar molecular pathways leading to common gene expression changes (Figure 4.3). These common changes (referred to as profiles, fingerprints, or signatures) can be used as specific and sensitive endpoints to identify and classify toxicants. These signatures fall under the definition of biomarkers of toxicity, since they serve as semiquantitative or quantitative indicators of the development or presence of a

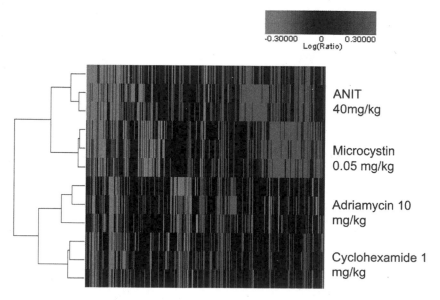

FIGURE 4.3 (See color insert following page 56.) Hierarchical clustering of gene profiles from the livers of rats treated for 3 days with various hepatotoxicants at toxic doses. Included in the heatmap are genes that were up- or down-regulated by a factor of ≥2 with a *p* value less than 0.01. Green indicates down-regulation, while red indicates up-regulation. Despite significant variability in responses observed with clinical pathology and histopathology, there is limited interindividual variability in gene expression profiles.

toxic change. The signature concept has been demonstrated in various types of diseases and toxic events [62–66]. In addition, through a global evaluation on the cell transcriptome, microarrays allow toxicologists to formulate novel hypotheses about mechanisms of action; these hypotheses can then be evaluated and confirmed in subsequent, appropriately designed experiments [63].

So far, most toxicogenomics studies have been conducted using either rat tissues or cell lines of rat or human origin. The major reason is that the rat is the most common small laboratory animal species for toxicology testing in the pharmaceutical industry. Other reasons include incomplete genome annotation in other species (such as the dog or monkey) and the lack of historical gene-expression data for these species. However, it is likely that, as this information becomes available, gene-expression profiling in other animal species will be more common.

The most important objective of toxicology is to identify toxic hazards and assess their risks for humans. Despite differences in mammalian genomes, responses to toxicants are often evolutionarily conserved between mammals. By identifying the affected cellular subsystems, gene profiling may indicate whether a toxic change is relevant to humans or predict how humans would react to a compound. Cross-species extrapolation facilitates prediction of toxicological reactions and risk assessment in humans [67].

4.5.2 OVERVIEW OF THE USE OF TOXICOGENOMICS IN PHARMACEUTICAL R&D

The goal of this chapter is to describe different applications of toxicogenomics using selected examples as illustrations. As mentioned earlier, toxicogenomics can be applied at different stages of drug development (Figure 4.1) to achieve the following:

- Prediction and characterization of the toxic attributes of experimental compounds from *in vitro* or *in vivo* studies
- Elucidation of the mechanism of toxicological changes
- Characterization of potential on-target toxicity issues associated with new therapeutic targets (i.e., toxicity related to primary pharmacology)

This chapter provides an overview of the use of toxicogenomics for predictive and mechanistic toxicology and for the characterization of on-target toxicity. Toxicogenomics is not intended to replace the traditional toxicology studies that are submitted as a part of a regulatory package; rather, it should be viewed as a complement to these studies. The major objective of toxicogenomics in predictive toxicology is to improve the productivity of pharmaceutical R&D by enabling the selection of development candidates with optimal toxicological profiles. Currently, candidate compounds are typically selected for development based on their physicochemical and pharmacological properties, with only limited toxicological characterization. Not surprisingly, this paradigm results in a high rate of failure in preclinical regulatory studies after significant resources have already been committed. By improving candidate selection through a more thorough early toxicological characterization in discovery, one could expect a significant reduction in the failure rate in early development. Another objective of toxicogenomics is to increase the value of regulatory preclinical studies by providing an improved understanding of the relevance of preclinical toxicological changes to humans, resulting in an overall better risk assessment. Toxicogenomics also offers a largely needed novel approach to identify additional toxicity biomarkers that could potentially be used to improve monitoring of adverse events in the clinics. This last aspect will not be covered in this chapter because of the lack of convincing examples in the literature. Unfortunately, it is still unclear how genomics-based biomarkers will be validated to become an integral part of regulated preclinical and clinical studies and of regulatory decision making. The validation of these biomarkers is complex and clearly context-specific with criteria that will be determined by their actual use.

4.5.3 PRACTICAL AND LOGISTIC ASPECTS OF TOXICOGENOMICS

Gene-expression profiling is a rapidly evolving, yet still immature, discipline. Consequently, various technical and practical issues have been identified and are being addressed by academic, governmental, and industry groups, often in the form of consortia or collaborations [68–72]. These technical issues range from external or internal control selection, probe design, platform variation and comparison, scanner performance characteristics, data normalization, or analysis software packages [69, 70, 73, 74]. Most current technical limitations are closely related to the instrumentation being used and are, consequently, rapidly evolving and becoming obsolete. For instance, early microarray platforms were associated with reproducibility and accuracy issues. The technology has since rapidly improved, and several recent studies have demonstrated the accuracy and reproducibility of large-scale gene-expression profiling [68, 71, 75]. For these reasons, this chapter will not address technical problems. Rather, we will focus here on critical issues related to the testing materials and design of studies.

Cell cultures are very homogeneous, and gene-expression changes induced by toxicants can be reliably reproduced under similar experimental conditions. Likewise, certain tissues, such as liver or heart, are sufficiently homogeneous in their phenotypes and transcriptomes to yield robust, consistent changes in gene expression following exposure to toxicants. In contrast, other tissues, such as brain or testis, are more heterogeneous and complex, and this represents a significant roadblock to the generation of consistent gene-expression profiles or to the identification of subtle gene-expression changes [76, 77]. This complexity had limited the use of toxicogenomics in these tissues. Gene profiling on single-cell populations is possible with the use of complementary technologies, such as laser capture microdissection (LCM) and RNA amplification protocols. Gene profiling of single cells may be applied to these complex tissues [78]. However, these technologies are resource-intensive and limit one's ability to fully understand major cell-cell interactions that may play a significant role in the pathogenesis of a toxicological change.

Toxicogenomics studies should be designed based on the question to be answered or the issue to be addressed. Hypothesis-driven experimental design will be reemphasized and illustrated as different applications are reviewed. However, several general principles apply to all toxicogenomics studies.

Dosing period: Gene profiles are transient and, therefore, timing is of critical importance. In particular, profiles that are more relevant to the mechanism of toxicological changes are typically those occurring before the phenotypic changes are fully developed. Consequently, in a study designed to address a toxic mechanism, it is generally more insightful to evaluate gene-expression profiles during the development of the toxic change, rather than when or after the toxic change is fully developed. This is in contrast to what is done in studies designed to predict toxic changes, where the performance characteristics of the predictive signatures dictate the duration of dosing. Hence, in our laboratory, since our predictive signatures have been validated for 3- or 5-day studies, we assess most exploratory compounds in 3- or 5-day rat studies. In several companies, toxicogenomic analysis is now typically integrated in most rat repeat-dose 2- or 4-week toxicology studies. The objective is to be proactive in case unexpected toxic changes occur in the studies or to enhance the knowledge in the gene-profiling area.

Doses: For any toxicology study, the use of appropriate doses is critical for the successful characterization or prediction of toxicological changes. In a mechanistic *in vivo* study, the dose is selected based on the best chance to consistently reproduce the toxicological change of interest. In contrast, in a predictive study, the performance characteristics (predictive power and accuracy for various dose levels) of the predictive signatures will dictate the dose to be used. For instance, in our laboratory, we developed our *in vivo* databases and signatures based on a low and a high dose. The low dose corresponds to an estimated pharmacological dose, while the high dose corresponds to a maximal tolerated dose for a 3- or 5-day study (defined by the maximal dose possible before there is evidence of clinical signs of toxicity or significant effect on body weight or food consumption).

4.5.4   TOXICOGENOMICS REFERENCE DATABASES

Gene-expression changes, when viewed in isolation, can be misinterpreted [79]. Examining gene-expression changes by looking at affected pathways, rather than by focusing on individual gene-expression changes, improves the accuracy of data interpretation, because the analysis of pathways increases the confidence that a change in expression of specific genes has biological implications [80]. However, drug-induced gene-expression profiles are best interpreted using appropriate reference databases containing gene-expression changes induced by a large repository of compounds or associated with specific toxic mechanisms [79, 80]. Indeed, gene-expression changes induced by toxicants reflect a large number of complex pharmacological, physiological, and biochemical processes [54, 81]. Consequently, not all gene-expression changes induced by toxicants are related to the toxicological process being investigated. In general, only the data (diversity of compounds and toxic changes, tissues, corroborative toxicological and pathologic changes, and gene-expression profiles) contained in large reference databases can accurately distinguish the gene-expression changes related to the toxicity from those that are adaptive, beneficial, or unrelated to the development of the toxicological change [82].

Toxicogenomics databases contain gene-expression profiles induced in various tissues following the treatment of the reference species (most often rats) with a variety of reference toxicants (known pharmaceutical agents and prototypical toxicants) and control compounds [64, 83–87]. To be useful, a toxicogenomics database needs to contain a set of reference data comparable to the gene-expression profiles being generated. Therefore, the reference compounds should cover a variety of toxic mechanisms and represent different structure–activity relationships. Evaluating several doses of the reference compounds (for instance, an efficacious dose and a maximum tolerated dose) is extremely useful to distinguish a pharmacological effect from a toxicological effect. Biological replicates are critical to establishing both the biological and technical variabilities. Fortunately, in species like rats, the interindividual variability of gene-expression changes is relatively small and typically lower than that seen with other toxicological endpoints. Therefore, three animals per group and per time point are usually sufficient to generate meaningful gene-expression profiles.

Ideally, gene-expression profiles generated in a particular species should be compared with those generated in the same species or even strain. Practically, this may not always be feasible because of a lack of reference data in some species or strains. This necessitates cross-species extrapolation. There have been significant improvements in the annotations of the genomes for the major preclinical species currently used in toxicology [67]. However, the mapping of orthologous genes is still approximate. Furthermore, not all species react similarly to a specific toxicant, and this difference in response also limits the ability of the researcher to extrapolate from one species to another. For instance, differences in responses to some toxicants between different strains of rats is a phenomenon well known to toxicologists. These differences appear to be of less concern with gene profiling, as the overall transcriptome response following exposure to toxicants has been shown to be very consistent across different rat strains [88].

Several publicly available commercial or proprietary databases exist for the analysis of gene-expression data sets [72, 82, 89–93]. Not all of these repositories are specific to toxicology. However, they still represent useful resources. A list of these databases with a brief description of their general attributes is provided in Table 4.1.

4.6 SPECIFIC APPLICATIONS OF TOXICOGENOMICS IN PHARMACEUTICAL R&D

4.6.1 PREDICTIVE TOXICOLOGY

The term *predictive toxicology* refers to the use of short-term assays to predict toxic changes that occur after a longer exposure. Predictive toxicology assays can be in the form of *in vitro* assays or of short-term *in vivo* studies. For simplicity, *in vitro* toxicogenomics will be covered in a different section. Following exposure to toxicants at relevant doses, transcriptional changes typically occur before the development of a toxic phenotype as assessed by traditional endpoints. Therefore, toxicogenomics provides a unique opportunity to identify compounds with toxic liabilities earlier

TABLE 4.1

Publicly Available Commercial or Proprietary Databases for the Analysis of Gene-Expression Data Sets

Database	Attributes	URL
Gene Expression Omnibus (GEO)	World's largest public repository Adherence to MIAME guidelines Toxicology data available Exploration, analysis, and visualization tools	http://www.ncbi.nlm.nih.gov/projects/geo/
ArrayExpress	Large public repository Adherence to MIAME guidelines Toxicology data available in Tox-MIAMExpress Expression data from normal human and mouse tissues	http://www.ebi.ac.uk/arrayexpress/
Chemical Effects in Biological Systems (CEBS)	Evolving public toxicogenomics repository from the National Institute of Environmental Health Sciences (NIEHS) National Center for Toxicogenomics (NCT) Designed to house data from complex studies having multiple data steams (genetic, proteomic, and metabonomics data) Exploration and analysis tools	http://cebs.niehs.nih.gov/
Environment, Drugs, and Gene Expression (EDGE)	Public toxicogenomics repository Standardized experimental conditions including standardized microarray platform Mostly focused on mouse liver microarray data Useful bioinformatics tools (clustering, BLAST searching, rank analysis, and classification tools)	http://edge.oncology.wisc.edu/
Symatlas	Product of the Genomics Institute of the Novartis Research Foundation Expression data from a large panel of normal human and mouse tissues or cell culture models	http://symatlas.gnf.org/SymAtlas/
DbZach System	Toxicogenomics database allowing data mining and full knowledge-based understanding of toxicological mechanisms Contains correlating clinical chemistry parameters and histopathologic data	http://dbzach.fst.msu.edu/

than with other methods. This is an area where toxicogenomics could most significantly impact the productivity of drug discovery [54, 94].

Predictive toxicology is applied at the lead selection and optimization stages concurrently with other assays used to assess druglike properties of molecules (pharmacological and physicochemical properties, ADME [absorption, distribution, metabolism, and excretion], and pharmacokinetics). However, current approaches are not yet fully amenable to such an early stage because of the low throughput and significant cost associated with gene-profiling technologies. It is likely that these factors will be rapidly addressed and that predictive toxicogenomics assays will become an integral part of the panel of assays used at the early stages of drug discovery.

Appropriate performance characteristics of the gene signatures are an absolute requirement to predict with confidence toxic changes with gene-expression profiling. It is beyond the scope of this chapter to provide an in-depth review of the methods used to generate predictive gene profiles. Several reviews are available on this topic [54, 82, 95, 96]. Development of predictive models to classify compounds as toxic or nontoxic requires the use of a variety of sophisticated statistical methodologies, computational algorithms, and a significant amount of biostatistical expertise. These tools use training sets that consist of a sufficiently large repository of gene-expression profiles encompassing both toxic and nontoxic compounds. Computational algorithms, such as linear discriminant analysis (LDA), Naïve Bayesian classifiers, artificial neural networks (ANN), and support vector machines (SVMs), are then used to classify unknown compounds [83, 85, 97, 98]. The prediction accuracy of the model is then typically estimated using a validation set independent from the training set [54, 66, 95].

Several proof-of-concept studies have demonstrated the feasibility and utility of predictive toxicogenomics. For instance, our laboratory has developed a quantitative approach to predict hepatotoxicity with a high degree of sensitivity and specificity. This approach relies on an internal toxicogenomics database containing microarray-generated liver gene expression profiles from 3- or 5-day rat toxicology studies. Using an ANN algorithm coupled with principal component analysis (PCA) for dimension reduction, we have developed a quantitative model to classify compounds according to a composite toxicity score. A forward validation step was conducted using additional compounds that were not part of the original database. This predictive hepatotoxicity assay is now routinely used to prioritize compounds using exploratory 3-day repeat-dose rat studies for various projects. In addition, in these short-term rat studies, we routinely collect additional tissues (heart, kidney, spleen, bone marrow, blood) for which predictive signatures have been developed. Gene-expression analysis is conducted after an initial evaluation of clinical pathology and histopathologic changes. These traditional toxicological measurements are then used to prioritize tissue evaluation by gene-expression analysis. Hence, should histological changes be identified in a tissue, this tissue would not be evaluated by gene-expression analysis, unless there is a need to understand the toxic mechanism.

Toxicogenomics has also been used to predict toxic changes in the kidney [66]. In this study, using a large commercial reference database, a predictive gene-expression signature of renal tubular toxicity was developed and shown to predict with good accuracy renal tubular changes that typically occur after longer exposures to the toxicants. A subsequent forward validation step included 21 compounds that were naïve and were not structurally related to the compounds in the training set. The signature correctly predicted renal tubular injury in 76% of the compound treatments (Figure 4.4).

Several investigators have also explored the possibility of developing predictive signatures for carcinogenic effects [84, 99, 100]. Carcinogenicity studies are

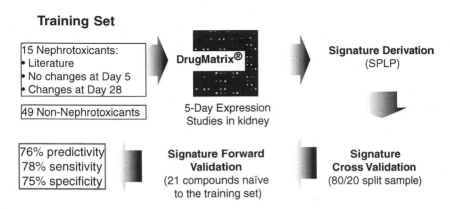

Training Set

FIGURE 4.4 Development of a predictive gene-expression signature for renal toxicity [67]. The training set consisted of renal gene-expression profiles from rats treated for 5 days with a variety of nephrotoxicants and non-nephrotoxicants. Using an SPLP algorithm, a signature was derived and cross-validated based on an 80/20 split sample approach. The signature was then further validated using a testing set composed of 21 compounds naïve to the training set. The signature had an overall predictive power of 76%.

required for the registration of new pharmaceutical agents intended for chronic or intermittent use over six months of duration [101]. However, assessment of the carcinogenic potential of compounds in the rodent bioassay is expensive, lengthy, and cannot be performed until late in a program. For these reasons, toxicogenomics applied in a predictive mode would allow for an earlier assessment of compounds. Three pioneering studies have successfully identified candidate molecular markers and pathways regulated by genotoxic and nongenotoxic hepatic carcinogens [84, 99, 100]. However, because of the limited size of the study set, it is impossible, at this point, to estimate the predictive value of these regulated genes and pathways.

The examples described above demonstrate that predictive gene signatures of relatively high accuracy can be generated once a sufficient repository of reference gene profiles is available. They also illustrate the scale and the amount of investment required to generate accurate predictive signatures. This suggests that continued refinement of these early signatures and development of new signatures will likely require a collaborative effort from the scientific community. Such combined effort of the broad scientific community would also provide a better estimate of their predictive power, robustness, and suitability for early toxicology testing.

4.6.2 IN VITRO TOXICOGENOMICS

Toxicologic assessment is ultimately conducted in animal studies. However, *in vivo* studies require large amounts of compound (in the gram range) and are expensive and time consuming. Consequently, they are not amenable to early and rapid characterization of toxicity for significant numbers of compounds. In contrast, *in vitro* systems have a higher throughput, are more cost-effective, and require much smaller amounts of compounds (in the milligram range). In addition, gene profiling using *in vitro* systems may identify biomarkers of toxicity in the form of gene sets that could be transferred to and investigated in preclinical or clinical studies to monitor possible toxic reactions. Finally, gene-expression studies in human cells, such as primary human hepatocytes, may, in some cases, be more relevant to the clinical situation or result in a better understanding of the relevance of changes, leading to a better assessment of safety risks for humans [102]. The major limitations of *in vitro* systems are their inability to recapitulate the overall complexity of the living organism and their short lifespan, which limits the potential to detect chronic effects [103]. In fact, understanding the real predictive value of *in vitro* systems has remained a major challenge for toxicologists for decades. Several studies have addressed the relationship of *in vitro* toxicogenomics results to *in vivo* data. It has been shown that different mechanisms of toxicity can be identified using gene profiles generated from *in vitro* systems. Consequently, the concept of predictive signatures is also relevant to *in vitro* systems [62, 104, 105].

For an *in vitro* toxicogenomics assay to have practical applications in drug discovery, gene-expression signatures of satisfactory accuracy need to be generated for several relevant toxicologic endpoints. To achieve this, cost-effective gene expression platforms with adequate throughput are required. To reach an appropriate throughput, it would be advantageous to reduce the number of genes being monitored. Ideally, one would want to rapidly evaluate compounds for several toxicologic endpoints in a simultaneous fashion. The selection of an appropriate dose also represents a critical issue. At this point, there is no clear consensus on the ideal dose for an *in vitro* toxicogenomics assessment. In our experience, appropriate characterization of toxicity requires the use of relatively high doses, sufficient to cause some degree of cytotoxicity. In our primary rat hepatocyte model, we typically characterize compounds at concentrations sufficient to cause 20% of cell death.

Most published *in vitro* toxicogenomics studies have evaluated rat-liver-derived *in vitro* systems for obvious reasons. First, the liver is a common, if not the most common, target organ of toxicity. Second, rat-liver cell preparations are traditionally used for *in vitro* toxicologic studies. Third, the use of hepatocytes offers the opportunity to assess the toxicity associated with certain metabolites without prior metabolic activation. Cell lines are easy to access, give reproducible results, and are very cost effective. However, liver cell lines are quite different from liver or primary hepatocytes in terms of function and phenotype [57]. In particular, liver cell lines express very low or undetectable levels of Phase I metabolizing enzymes [57]. Consequently, evaluating compounds in cell lines limits the detection of toxic changes related to the formation of metabolites. Isolated hepatocytes are not identical to, but sufficiently close to, intact livers in terms of gene-expression analysis [57, 106]. Most importantly, they maintain their metabolizing capabilities in short-term cultures [107, 108]. In our laboratory, we routinely use primary rat hepatocytes cultured on collagen to characterize compounds by gene profiling. We have profiled a large number of diverse compounds, thereby generating an internal database that has allowed us to develop predictive signatures for several toxicological endpoints, such as phospholipidosis, DNA damage, peroxisome proliferation, or glutathione depletion. For instance, our laboratory has reported results from a profiling study of 15 well-characterized hepatotoxicants in primary rat hepatocytes [104]. Compounds with similar mechanisms of toxicity resulted in similar expression profiles, and, using unsupervised hierarchical clustering, could clearly be distinguished from other agents. This study also demonstrated a significant correlation between the genes regulated *in vivo* and *in vitro* for some toxicants, such as aryl hydrocarbon receptor (AhR) ligands. We also have recently reported two predictive signatures for toxicologically relevant endpoints [62]. Similar results have been reported by others for rat hepatocytes and hepatic cell lines [58, 109–111].

Toxicogenomics has also been applied in non-liver-derived cell systems. In particular, gene-expression analysis has been used to evaluate genotoxic mechanisms. The current *in vitro* genotoxicity assays using mammalian cells (mammalian mutation and/or chromosomal damage assays) have low specificity and may not be relevant to the *in vivo* situation [59, 112, 113]. Assays that differentiate DNA-reactive versus non-DNA-reactive mechanisms of genotoxicity improve the risk assessment of positive findings in the *in vitro* mammalian cell-based assays. Several gene-profiling studies have demonstrated differences in gene-expression profiles between DNA-damaging and non-DNA-damaging compounds [59, 112, 114]. This suggests that, although toxicogenomics will not replace the current standard genotoxicity assays for hazard identification, it can serve as a useful follow-up experimental approach to evaluate compounds with positive findings in these standard assays.

A significant effort is needed for *in vitro* toxicogenomics to become part of the battery of routine assays used at the lead selection and lead optimization stages of drug discovery. However, this area is rapidly improving, and several companies have committed to the use of gene-expression-based *in vitro* assays for compound characterization and prioritization. For instance, our laboratory has already developed several specific predictive signatures in the rat hepatocyte model, and *in vitro* toxicogenomics is rapidly being integrated in the standard testing funnel for most of our discovery projects [54, 62, 115].

4.6.3 Mechanistic Toxicology

Gene profiling has so far been most useful when elucidating the mechanism of a specific toxicity. Toxic changes are commonly identified in preclinical studies, and obviously not all toxic changes are worth investigating [54]. Typically, at this stage, a significant amount of resources has already been invested, and the value of specific studies becomes more apparent. For instance, if a toxicological change is considered development-limiting and no backup compounds are available, developing a mechanistic understanding of a toxic change may facilitate the development of appropriate counterscreens for selecting backup compounds unlikely to induce the same toxic change.

Early studies have shown that changes in gene expression can help formulate mechanistic hypotheses. This can be achieved by comparing expression profiles with those present in a database [54, 64, 116]. For instance, our laboratory has investigated the hepatic effects of A-277249, a thienopyridine inhibitor of NF-κB-mediated expression of cellular adhesion molecules [117]. This compound induced hepatic toxicity in rats in a repeat-dose toxicity study. To investigate the mechanism of this hepatotoxicity, a 3-day repeat-dose rat toxicity study was conducted, and gene-expression profiles from livers were generated. Using our proprietary gene-expression database of known hepatotoxicants, we applied agglomerative hierarchical cluster analysis to demonstrate that A-277249 had a gene-expression profile similar to activators of the aryl hydrocarbon nuclear receptor (AhR). This implies that A-277249 hepatic changes were, at least in part, mediated by the AhR either directly or through effects on NF-κB [117].

Two recent studies used gene-expression profiling to develop a mechanistic understanding of the gastrointestinal toxicity induced by functional κ-secretase inhibitors (FGSIs) in rats. The FGSIs have been developed as potential therapeutic agents for Alzheimer's disease [118, 119]. They can block the cleavage of several transmembrane proteins, including the cell fate regulator Notch-1, which plays an important role in the differentiation of the immune system and the gastrointestinal tract. Rats treated with several FGSIs developed a mucoid enteropathy related to ileal goblet cell hyperplasia [118, 119]. Microarray analysis of the duodenum and ileum of FGSIs-treated rats identified changes in the expression of several genes, and these changes confirmed that perturbation in Notch signaling was the mechanism for this unique enteropathy. These gene-expression studies also identified that the gene encoding the serine protease adipsin was significantly up-regulated following treatment with FGSIs. The investigators confirmed this finding by demonstrating elevated levels of the adipsin protein in gastrointestinal contents and feces of FGSIs-treated rats, as well as increased numbers of ileal enterocytes expressing adipsin by immunohistochemistry. These investigations led to the identification of adipsin as a potential noninvasive biomarker of FGSIs-induced gastrointestinal toxicity. Of particular interest is the fact that these studies were conducted almost simultaneously in different institutions, demonstrating the reproducibility of gene-profiling results across laboratories.

4.6.4 TOXICOGENOMICS AND TARGET-RELATED TOXICITY

Understanding the potential toxicity associated with modulation of the activity of a particular therapeutic target (or what is typically referred to as target-related toxicity, mechanism-based toxicity, or on-target toxicity) can be useful in prioritizing targets that are most likely to succeed in a discovery portfolio. Clearly, most of the known genomics-derived targets play major roles in normal cellular function, and consequently modulation of their activity can lead to toxic changes. Therefore, developing a good understanding of the potential safety liabilities associated with a target should be viewed as a critical phase of target druggability assessment. Gene-expression profiling provides an improved understanding of the biology of therapeutic targets and can help identify target-related safety liabilities. Ideally, this assessment should be conducted at the earliest stages of the discovery process, namely, the target identification/validation stage. One approach is to investigate target expression in normal tissues from both preclinical species and human beings in order to proactively identify tissues more likely to be affected by toxic changes [94, 120–122]. The availability of complete tissue repositories or banks is necessary to generate these tissue expression maps. Target expression can be evaluated at the level of RNA expression, protein expression, or enzymatic activity, or using receptor-binding assays.

In our experience, albeit useful, target expression in normal tissues is usually not sufficient to predict on-target toxicity. This is best achieved using appropriate tool compounds (i.e., compounds with adequate pharmacological activity against the target and sufficient selectivity). To control for toxicity related to a chemical class, one should use several tool compounds from different chemical classes and, if available, inactive compounds with close structural similarities (such as inactive enantiomers, for instance). In our laboratory, we typically conduct 3-day repeat-dose rat studies using carefully selected tool compounds. Doses are selected to result in exposures sufficient to achieve efficacy, but also in higher exposures, so that target-related safety margins can be estimated. Tissues for gene-expression profiling are selected and prioritized based on the biology of the target and a prior evaluation of clinical pathology and histopathology changes. Gene-expression changes are then evaluated in the context of available reference databases as well with as our battery of gene-expression signatures.

Tools used for target validation (antibodies, genetically engineered mice, or technologies to modulate mRNA expression levels, such as antisense oligonucleotides, ribozymes, or siRNA) can theoretically be used to predict on-target toxicity issues [123–125]. In particular, siRNA has gained popularity in the last few years. Our personal experience with siRNA in *in vitro* systems has been mixed, suggesting that this tool needs to be better understood before it can be applied to characterize on-target toxicity. In particular, gene silencing approaches induce varying degrees of mRNA and protein down-regulation, which may not mimic the pharmacological inhibition necessary to achieve efficacy. RNA interference (RNAi) has successfully been used *in vivo* [126, 127]. However, the costs of siRNA limits these *in vivo* studies to mice, a species for which no robust toxicogenomics databases are available, and the delivery techniques induce mild adverse effects that complicate the interpretation of gene profiles. Clearly, the current experience with *in vivo* RNAi studies is too limited to reliably evaluate whether gene silencing can be useful in predicting on-target toxicity. However, this approach is conceptually very promising.

4.6.5 TOXICOGENOMICS IN REGULATORY SUBMISSIONS

The FDA recognized the potential of gene-expression profiling to improve the safety assessment of new chemical entities and has issued guidance in March 2005 for the regulatory submission of pharmacogenomics data (http://www.fda.gov/cder/guidance). This guidance reflects a laudable effort of the FDA to promote the use of genomic technologies in drug development and is also designed to enhance the agency's knowledge of these emerging technologies. In finalizing this guideline, the FDA has openly cooperated with the various stakeholders and organized appropriate forums to focus on the major issues and principles that the document should cover. This guidance represents an important stepping stone toward the development of genomics-based drugs and the effective use of genomics-based safety data. Importantly, this guidance provided reassurance to companies that early-stage toxicogenomics experiments would not bring negative regulatory consequences, an important aspect for the wider acceptance of this new technology in the traditionally conservative environment of drug safety evaluation.

The guidance is an effort to clarify the FDA's policy on the use of pharmacogenomics data in the drug application review process. It covers the application of genomics concepts and technologies to nonclinical and clinical pharmacology and clinical studies. In particular, guidelines are provided regarding the requirements for submission of pharmacogenomics data, the format and procedure for data submission, and how the data will be used in regulatory decision making. Briefly, submission is required for:

- Data used for decision making within a specific trial
- Evidence in support of scientific arguments about the mechanisms of action, dose selection, safety, or efficacy
- Data that will support registration or labeling language
- Results generated for previously validated biomarkers

This guidance indicates that the FDA is open to and expects submission of gene-profiling data that were generated to support scientific contentions related to toxicity.
    The guidance defines pharmacogenomic tests as follows:

An assay intended to study interindividual variations in whole-genome or candidate gene, single-nucleotide polymorphism (SNP) maps, haplotype markers, or alterations in gene expression or inactivation that may be correlated with pharmacological and therapeutic responses. In some cases, the pattern or profile of change is the relevant biomarker, rather than changes in individual markers.

This implies that gene-expression datasets could ultimately be recognized by the agency as validated biomarkers. The guidance also defines "valid biomarkers" and distinguishes between "known valid biomarkers" and "probable valid biomarkers." Valid biomarkers are measured in an analytical test system with well-established performance characteristics, when an established scientific framework or body of evidence exists to help understand the significance of the test results. For a known valid biomarker, a widespread agreement exists in the medical and scientific communities about the significance of the results. In contrast, for a probable valid biomarker, there is no widespread agreement, but only a scientific framework or body of evidence sufficient to elucidate the significance of the test results. An example would be a biomarker developed by a sponsor and not available for public scientific scrutiny or for independent verification. Albeit quite confusing and debatable, this distinction indicates the enormous amount of work and improvement that will be needed in the future to make gene-expression data suitable for regulatory decision making. This obviously includes an improved scientific framework for data interpretation through the use of larger, more complete reference databases, but also improved quality control of laboratory procedures, a better understanding of the comparability of different platforms, and a well-defined process for biomarker validation. Furthermore, it is still unclear how genomics-based biomarkers will be validated to become an integral part of regulatory decision making. The validation of these biomarkers is complex and clearly context-specific with criteria that will depend on their actual use [128]. For these reasons, it is clear that incorporation of validated genomic biomarkers into regulated studies will require some time.

For investigators using genomics technologies as an experimental tool, an interesting aspect of the guidance is the concept of "Voluntary Genomic Data Submission" (VGDS). The VGDS concept reflects the recognition by the agency that, currently, most gene-profiling data are of exploratory or research nature, and would therefore not be required for submission. However, to be prepared to appropriately evaluate future submissions, FDA scientists and study sponsors need to develop an understanding of a variety of relevant scientific and procedural issues. The VGDS provides the material necessary to build this understanding, and is reviewed by a cross-center Interdisciplinary Pharmacogenomic Review Group (IPRG). For more information, the reader is referred to the FDA Web site that reviews the FAQs (frequently asked questions) regarding VGDS (http://www.fda.gov/cder/genomics/FAQ.htm). Briefly, the IPRG review of a VGDS is a scientific discussion in parallel to a regulatory context. All VGDS data are protected from disclosure either outside the FDA or to review divisions, and are routed directly to the IPRG and stored on a separate secure server. These data are not distributed outside the IPRG without prior agreement with the sponsor and are not to be used for regulatory decision making. The concept of VGDS has in general been well received by the pharmaceutical industry. Voluntary submissions allow sponsors to familiarize FDA scientists with genomics data and their interpretation, and, at the same time, to learn the regulatory decision-making process and expectations involving genomics data. Ultimately, this could prevent delays in the future submissions containing required genomics data. So far, more than a dozen formal submissions have been made.

This guidance indicates that the FDA is expecting genomics data to become an integrated part of the risk assessment of pharmaceuticals. It is, however, difficult at this point to objectively predict the role that a rapidly evolving technology will have in regulatory decision making. When the concept of toxicogenomics first emerged, expectations were high and to some extent unrealistic. The technology and the analytical tools have since considerably improved in terms of robustness, reproducibility, and accuracy. However, genomics is still an immature approach to evaluate toxicology, and, as stated earlier, toxicogenomics should not be viewed as a replacement for traditional toxicology studies that are part of a regulatory package. Rather, toxicogenomics is a complement to these studies. When used as a predictive tool, it can be used to make better compound selection decisions. In this situation, regulatory submission of toxicogenomics data is not expected. When used as a tool to elucidate mechanisms of toxicity, toxicogenomics will result in data that could be used to support scientific arguments about the mechanism of action, and in these situations, regulatory submission will be required.

4.7 CONCLUSION

The decoding of the human genome sequence has led to a revolution in the pharmaceutical industry [14, 129]. Drug development has undergone a major paradigm change by shifting from a trial-and-error approach to a systematic hypothesis-driven strategy based on target identification and validation, followed by selection and optimization of compounds that inhibit the target with minimal side effects. Today, the early stages of the drug development process—target identification and validation—

almost inevitably involve gene-profiling experiments. For example, target identification in cancer often involves large-scale genomic screens for genes or their products that alter cell proliferation and survival. Target validation often involves modulation of the target activity followed by gene profiling to identify the pathways affected by target knockdown. Later in the drug discovery process, gene profiling plays a growing role in compound assessment. To address the high rate of failure due to safety and efficacy, the pharmaceutical industry needs to more accurately estimate efficacy and identify toxicity hazards earlier in the development process. As illustrated in this chapter, gene-expression profiling represents an additional tool to meet this objective.

A compelling amount of data shows that when used in the proper context, genomics technologies can significantly impact the productivity of pharmaceutical R&D. A particular challenge is the proper integration of these technologies within pharmaceutical R&D units. Successful genomics endeavors require broad and diverse expertise, often achieved only in the form of multidisciplinary teams composed not only of biologists, chemists, or toxicologists, but also of bioinformaticians and biostatisticians. Most major pharmaceutical companies have committed to significant investments, but so far, one has to acknowledge that, in contrast to its use in efficacy evaluation, gene profiling in toxicologic evaluation has still not been fully integrated in many organizations. This may reflect the current shortage of experts with a pragmatic vision of the future use of this technology, as well as the traditionally conservative nature of toxicology departments in the pharmaceutical industry. Nevertheless, if the trend seen in the last few years continues, it is realistic to predict that toxicogenomics will also have a growing strategic role in the risk assessment of new chemical entities in the pharmaceutical industry.

This overview of the use of gene profiling at different stages of drug discovery demonstrates that the success of the new drug development paradigm is dependent on the genomic data. Therefore, we believe that there is no alternative to continuing and expanding the use of the gene-profiling technology in drug discovery and development.

REFERENCES

1. Rawlings, M. D. 2004. Cutting the cost of drug development? *Nat. Rev. Drug Discov.* 3: 360.
2. Service, R. F. 2004. Surviving the blockbuster syndrome. *Science* 303: 1796.
3. Kola, I., and Landis, J. 2004. Can the pharmaceutical industry reduce attrition rates? *Nat. Rev. Drug Discov.* 3: 711.
4. Pollack, J. R., Sorlie, T., Perou, C. M., Rees, C. A., et al. 2002. Microarray analysis reveals a major direct role of DNA copy number alteration in the transcriptional program of human breast tumors. *Proc. Natl. Acad. Sci. USA* 99: 12963.
5. Linn, S. C., West, R. B., Pollack, J. R., Zhu, S., et al. 2003. Gene expression patterns and gene copy number changes in dermatofibrosarcoma protuberans. *Am. J. Pathol.* 163: 2383.
6. Heidenblad, M., Lindgren, D., Veltman, J. A., Jonson, T., et al. 2005. Microarray analyses reveal strong influence of DNA copy number alterations on the transcriptional patterns in pancreatic cancer: implications for the interpretation of genomic amplifications. *Oncogene* 24: 1794.

7. Lewin, D. A., and Weiner, M. P. 2004. Molecular biomarkers in drug development. *Drug Discov. Today* 9: 976.
8. Frank, R., and Hargreaves, R. 2003. Clinical biomarkers in drug discovery and development. *Nat. Rev. Drug Discov.* 2: 566.
9. Carter, P., Presta, L., Gorman, C. M., Ridgway, J. B., et al. 1992. Humanization of an anti-p185HER2 antibody for human cancer therapy. *Proc. Natl. Acad. Sci. USA* 89: 4285.
10. Cobleigh, M. A., Vogel, C. L., Tripathy, D., Robert, N. J., et al. 1999. Multinational study of the efficacy and safety of humanized anti-HER2 monoclonal antibody in women who have HER2-overexpressing metastatic breast cancer that has progressed after chemotherapy for metastatic disease. *J. Clin. Oncol.* 17: 2639.
11. Vogel, C. L., Cobleigh, M. A., Tripathy, D., Gutheil, J. C., et al. 2002. Efficacy and safety of trastuzumab as a single agent in first-line treatment of HER2-overexpressing metastatic breast cancer. *J. Clin. Oncol.* 20: 719.
12. Orth, A. P., Batalov, S., Perrone, M., and Chanda, S. K. 2004. The promise of genomics to identify novel therapeutic targets. *Expert Opin. Ther. Targets* 8: 587.
13. Pollack, J. R., Perou, C. M., Alizadeh, A. A., Eisen, M. B., et al. 1999. Genome-wide analysis of DNA copy-number changes using cDNA microarrays. *Nat. Genet.* 23: 41.
14. Lander, E. S., Linton, L. M., Birren, B., Nusbaum, C., et al. 2001. Initial sequencing and analysis of the human genome. *Nature* 409: 860.
15. Broder, S., and Venter, J. C. 2000. Sequencing the entire genomes of free-living organisms: the foundation of pharmacology in the new millennium. *Annu. Rev. Pharmacol. Toxicol.* 40: 97.
16. Golub, T. R., Slonim, D. K., Tamayo, P., Hvard, C., et al. 1999. Molecular classification of cancer: class discovery and class prediction by gene expression monitoring. *Science* 286: 531.
17. Abu-Elheiga, L., Matzuk, M. M., Kordari, P., Oh, W., et al. 2005. Mutant mice lacking acetyl-CoA carboxylase 1 are embryonically lethal. *Proc. Natl. Acad. Sci. USA* 102: 12011.
18. Yeoh, E. J., Ross, M. E., Shurtleff, S. A., Williams, W. K., et al. 2002. Classification, subtype discovery, and prediction of outcome in pediatric acute lymphoblastic leukemia by gene expression profiling. *Cancer Cell* 1: 133.
19. Armstrong, S. A., Staunton, J. E., Silverman, L. B., Pieters, R., et al. 2002. MLL translocations specify a distinct gene expression profile that distinguishes a unique leukemia. *Nat. Genet.* 30: 41.
20. Armstrong, S. A., Kung, A. L., Mabon, M. E., Silverman, L. B., et al. 2003. Inhibition of FLT3 in MLL: validation of a therapeutic target identified by gene expression based classification. *Cancer Cell* 3: 173.
21. MacDonald, T. J., Brown, K. M., LaFleur, B., Peterson, K., et al. 2001. Expression profiling of medulloblastoma: PDGFRA and the RAS/MAPK pathway as therapeutic targets for metastatic disease. *Nat. Genet.* 29: 143.
22. Dhanasekaran, S. M., Barrette, T. R., Ghosh, D., Shah, R., et al. 2001. Delineation of prognostic biomarkers in prostate cancer. *Nature* 412: 822.
23. Magee, J. A., Araki, T., Patil, S., Ehrig, T., et al. 2001. Expression profiling reveals hepsin overexpression in prostate cancer. *Cancer Res.* 61: 5692.
24. Fromont, G., Chene, L., Vidaud, M., Vallancien, G., et al. 2005. Differential expression of 37 selected genes in hormone-refractory prostate cancer using quantitative taqman real-time RT-PCR. *Int. J. Cancer* 114: 174.
25. Doniger, S. W., Salomonis, N., Dahlquist, K. D., Vranizan, K., et al. 2003. MAPPFinder: using Gene Ontology and GenMAPP to create a global gene-expression profile from microarray data. *Genome Biol.* 4: R7.

26. Volinia, S., Evangelisti, R., Francioso, F., Arcelli, D., et al. 2004. GOAL: automated Gene Ontology analysis of expression profiles. *Nucleic Acids Res.* 32: 492.

27. Al-Shahrour, F., Minguez, P., Vaquerizas, J. M., Conde, L., et al. 2005. BABELO-MICS: a suite of web tools for functional annotation and analysis of groups of genes in high-throughput experiments. *Nucleic Acids Res.* 33: 460.

28. Mlecnik, B., Scheideler, M., Hackl, H., Hartler, J., et al. 2005. PathwayExplorer: web service for visualizing high-throughput expression data on biological pathways. *Nucleic Acids Res.* 33: 633.

29. Nikolsky, Y., Ekins, S., Nikolskaya, T., and Bugrim, A. 2005. A novel method for generation of signature networks as biomarkers from complex high throughput data. *Toxicol. Lett.* 158: 20.

30. Elbashir, S. M., Lendeckel, W., and Tuschl, T. 2001. RNA interference is mediated by 21- and 22-nucleotide RNAs. *Genes Dev.* 15: 188.

31. Elbashir, S. M., Harborth, J., Lendeckel, W., Yalcin, A., et al. 2001. Duplexes of 21-nucleotide RNAs mediate RNA interference in cultured mammalian cells. *Nature* 411: 494.

32. Sarang, S. S., Yoshida, T., Cadet, R., Valeras, A. S., et al. 2002. Discovery of molecular mechanisms of neuroprotection using cell-based bioassays and oligonucleotide arrays. *Physiol. Genomics* 11: 45.

33. Svaren, J., Ehrig, T., Abdulkadir, S. A., Ehrengruber, M. U., et al. 2000. EGR1 target genes in prostate carcinoma cells identified by microarray analysis. *J. Biol. Chem.* 275: 38524.

34. Huang, E., Ishida, S., Pittman, J., Dressman, H., et al. 2003. Gene expression phenotypic models that predict the activity of oncogenic pathways. *Nat. Genet.* 34: 226.

35. Cho, Y. S., Kim, M. K., Cheadle, C., Neary, C., et al. 2001. Antisense DNAs as multi-site genomic modulators identified by DNA microarray. *Proc. Natl. Acad. Sci. USA* 98: 9819.

36. Semizarov, D., Kroeger, P., and Fesik, S. 2004. siRNA-mediated gene silencing: a global genome view. *Nucleic Acids Res.* 32: 3836.

37. Ziegelbauer, J., Wei, J., and Tjian, R. 2004. Myc-interacting protein 1 target gene profile: a link to microtubules, extracellular signal-regulated kinase, and cell growth. *Proc. Natl. Acad. Sci. USA* 101: 458.

38. Glaser, K. B., Staver, M. J., Waring, J. F., Stender, J., et al. 2003. Gene expression profiling of multiple histone deacetylase (HDAC) inhibitors: defining a common gene set produced by HDAC inhibition in T24 and MDA carcinoma cell lines. *Mol. Cancer Ther.* 2: 151.

39. Bottone, F. G., Jr., Martinez, J. M., Collins, J. B., Afshari, C. A., et al. 2003. Gene modulation by the cyclooxygenase inhibitor, sulindac sulfide, in human colorectal carcinoma cells: possible link to apoptosis. *J. Biol. Chem.* 278: 25790.

40. Clarke, P. A., te Poele, R., and Workman, P. 2004. Gene expression microarray technologies in the development of new therapeutic agents. *Eur. J. Cancer* 40: 2560.

41. Gunther, E. C., Stone, D. J., Gerwien, R. W., Bento, P., et al. 2003. Prediction of clinical drug efficacy by classification of drug-induced genomic expression profiles in vitro. *Proc. Natl. Acad. Sci. USA* 100: 9608.

42. Biomarkers Definitions Working Group. 2001. Biomarkers and surrogate endpoints: preferred definitions and conceptual framework. *Clin. Pharmacol. Ther.* 69: 89.

43. Ross, D. T., Scherf, U., Eisen, M. B., Perou, C. M., et al. 2000. Systematic variation in gene expression patterns in human cancer cell lines. *Nat. Genet.* 24: 227.

44. Scherf, U., Ross, D. T., Waltham, M., Smith, L. H., et al. 2000. A gene expression database for the molecular pharmacology of cancer. *Nat. Genet.* 24: 236.

45. Staunton, J. E., Slonim, D. K., Coller, H. A., Tamayo, P., et al. 2001. Chemosensitivity prediction by transcriptional profiling. *Proc. Natl. Acad. Sci. USA* 98: 10787.

46. Scherf, U., Ross, D. T., Waltham, M., Smith, L.,H., et al. 2000. A gene expression database for the molecular pharmacology of cancer. *Nat. Genet.* 24: 236.
47. Zembutsu, H., Ohnishi, Y., Tsunoda, T., Furukawa, Y., et al. 2002. Genome-wide cDNA microarray screening to correlate gene expression profiles with sensitivity of 85 human cancer xenografts to anticancer drugs. *Cancer Res.* 62: 518.
48. Zembutsu, H., Ohnishi, Y., Daigo, Y., Katagiri, T., et al. 2003. Gene-expression profiles of human tumor xenografts in nude mice treated orally with the EGFR tyrosine kinase inhibitor ZD1839. *Int. J. Oncol.* 23: 29.
49. Leung, S. Y., Chen, X., Chu, K. M., Yuen, S. T., et al. 2002. Phospholipase A2 group IIA expression in gastric adenocarcinoma is associated with prolonged survival and less frequent metastasis. *Proc. Natl. Acad. Sci. USA* 99: 16203.
50. Bohen, S. P., Troyanskaya, O. G., Alter, O., Warnke, R., et al. 2003. Variation in gene expression patterns in follicular lymphoma and the response to rituximab. *Proc. Natl. Acad. Sci. USA* 100: 1926.
51. Chen, X., Leung, S. Y., Yuen, S. T., Chu, K. M., et al. 2003. Variation in gene expression patterns in human gastric cancers. *Mol. Biol. Cell* 14: 3208.
52. Oshita, F., Ikehara, M., Sekiyama, A., Hamanaka, N., et al. 2004. Genomic-wide cDNA microarray screening to correlate gene expression profile with chemoresistance in patients with advanced lung cancer. *J. Exp. Ther. Oncol.* 4: 155.
53. Takata, R., Katagiri, T., Kanehira, M., Tsunoda, T., et al. 2005. Predicting response to methotrexate, vinblastine, doxorubicin, and cisplatin neoadjuvant chemotherapy for bladder cancers through genome-wide gene expression profiling. *Clin. Cancer Res.* 11: 2625.
54. Yang, Y., Blomme, E. A., and Waring, J. F. 2004. Toxicogenomics in drug discovery: from preclinical studies to clinical trials. *Chem. Biol. Interact.* 150: 71.
55. Schena, M., Shalon, D., Davis, R. W., and Brown, P. O. 1995. Quantitative monitoring of gene expression patterns with a complementary DNA microarray. *Science* 270: 467.
56. Nuwaysir, E. F., Bittner, M., Trent, J., Barrett, J. C., et al. 1999. Microarrays and toxicology: the advent of toxicogenomics. *Mol. Carcinog.* 24: 153.
57. Boess, F., Kamber, M., Romer, S., Gasser, R., et al. 2003. Gene expression in two hepatic cell lines, cultured primary hepatocytes, and liver slices compared to the in vivo liver gene expression in rats: possible implications for toxicogenomics use of in vitro systems. *Toxicol. Sci.* 73: 386.
58. Hong, Y., Muller, U. R., and Lai, F. 2003. Discriminating two classes of toxicants through expression analysis of HepG2 cells with DNA arrays. *Toxicol. In Vitro* 17: 85.
59. Newton, R. K., Aardema, M., and Aubrecht, J. 2004. The utility of DNA microarrays for characterizing genotoxicity. *Environ. Health Perspect.* 112: 420.
60. U.S. Department of Health and Human Services, FDA. 2004. Innovation or stagnation? Challenge and opportunity on the critical path to new medical products.
61. MacNeil, J. S. 2005. Genomics goes downstream. *Genome Technol.* 54: 24.
62. Yang, Y., Abel, S. J., Ciurlionis, R., and Waring, J. F. 2006. Development of a toxicogenomics in vitro assay for the efficient characterization of compounds. *Pharmacogenomics* 7: 177.
63. Segal, E., Friedman, N., Kaminski, N., Regev, A., et al. 2005. From signatures to models: understanding cancer using microarrays. *Nat. Genet.* 37: S38.
64. Hamadeh, H. K., Bushel, P. R., Jayadev, S., Martin, K., et al. 2002. Gene expression analysis reveals chemical-specific profiles. *Toxicol. Sci.* 67: 219.
65. Lee, J., Richburg, J. H., Shipp, E. B., Meistrich, M. L., et al. 1999. The Fas system, a regulator of testicular germ cell apoptosis, is differentially up-regulated in Sertoli cell versus germ cell injury of the testis. *Endocrinology* 140: 852.

66. Fielden, M. R., Eynon, B. P., Natsoulis, G., Jarnagin, K., et al. 2005. A gene expression signature that predicts the future onset of drug-induced renal tubular toxicity. *Toxicol. Pathol.* 33: 675.

67. Thomas, R. S., Rank, D. R., Penn, S. G., Zastrow, G. M., et al. 2002. Applications of genomics to toxicology research. *Environ. Health Perspect.* 110: 919–23.

68. Chu, T. M., Deng, S., Wolfinger, R., Paules, R. S., et al. 2004. Cross-site comparison of gene expression data reveals high similarity. *Environ. Health Perspect.* 112: 449.

69. Shi, L., Tong, W., Goodsaid, F., Frueh, F. W., et al. 2004. QA/QC: challenges and pitfalls facing the microarray community and regulatory agencies. *Expert. Rev. Mol. Diagn.* 4: 761.

70. Shi, L., Tong, W., Fang, H., Scherf, U., et al. 2005. Cross-platform comparability of microarray technology: intra-platform consistency and appropriate data analysis procedures are essential. *BMC Bioinformatics* 6: S12.

71. Waring, J. F., Ulrich, R. G., Flint, N., Morfitt, D., et al. 2004. Interlaboratory evaluation of rat hepatic gene expression changes induced by methapyrilene. *Environ. Health Perspect.* 112: 439.

72. Mattes, W. B., Pettit, S. D., Sansone, S. A., Bushel, P. R., et al. 2004. Database development in toxicogenomics: issues and efforts. *Environ. Health Perspect.* 112: 495.

73. Shi, L., Tong, W., Su, Z., Han, T., et al. 2005. Microarray scanner calibration curves: characteristics and implications. *BMC Bioinformatics* 6: S11.

74. Thompson, K. L., Rosenzweig, B. A., Pine, P. S., Retief, J., et al. 2005. Use of a mixed tissue RNA design for performance assessments on multiple microarray formats. *Nucleic Acids Res.* 33: e187.

75. Baker, V. A., Harries, H. M., Waring, J. F., Duggan, C. M., et al. 2004. Clofibrate-induced gene expression changes in rat liver: a cross-laboratory analysis using membrane cDNA arrays. *Environ. Health Perspect.* 112: 428.

76. Mirnics, K., and Pevsner, J. 2004. Progress in the use of microarray technology to study the neurobiology of disease. *Nat. Neurosci.* 7: 434.

77. Galvin, J. E., and Ginsberg, S. D. 2004. Expression profiling and pharmacotherapeutic development in the central nervous system. *Alzheimer Dis. Assoc. Disord.* 18: 264.

78. Todd, R., Lingen, M. W., and Kuo, W. P. 2002. Gene expression profiling using laser capture microdissection. *Expert Rev. Mol. Diagn.* 2: 497.

79. Kolaja, K., and Fielden, M. 2004. The impact of toxicogenomics on preclinical development: from promises to realized value to regulatory implications. *Preclinica* 2: 122.

80. Fielden, M. R., Pearson, C., Brennan, R., and Kolaja, K. L. 2005. Preclinical drug safety analysis by chemogenomic profiling in the liver. *Am. J. Pharmacogenomics* 5: 161.

81. Guerreiro, N., Staedtler, F., Grenet, O., Kehren, J., et al. 2003. Toxicogenomics in drug development. *Toxicol. Pathol.* 31: 471.

82. Ganter, B., Tugendreich, S., Pearson, C. I., Ayanoglu, E., et al. 2005. Development of a large-scale chemogenomics database to improve drug candidate selection and to understand mechanisms of chemical toxicity and action. *J. Biotechnol.* 119: 219.

83. Bushel, P. R., Hamadeh, H. K., Bennett, L., Green, J., et al. 2002. Computational selection of distinct class- and subclass-specific gene expression signatures. *J. Biomed. Inform.* 35: 160.

84. Ellinger-Ziegelbauer, H., Stuart, B., Wahle, B., Bomann, W., et al. 2004. Characteristic expression profiles induced by genotoxic carcinogens in rat liver. *Toxicol. Sci.* 77: 19.

85. Thomas, R. S., Rank, D. R., Penn, S. G., Zastrow, G. M., et al. 2001. Identification of toxicologically predictive gene sets using cDNA microarrays. *Mol. Pharmacol.* 60: 1189.

86. Waring, J. F., Jolly, R. A., Ciurlionis, R., Lum, P. Y., et al. 2001. Clustering of hepato-toxins based on mechanism of toxicity using gene expression profiles. *Toxicol. Appl. Pharmacol.* 175: 28.

87. Waring, J. F., Cavet, G., Jolly, R. A., McDowell, J., et al. 2003. Development of a DNA microarray for toxicology based on hepatotoxin-regulated sequences. *Environ. Health. Perspect.* 111: 863.

88. Luhe, A., Suter, L., Ruepp, S., Singer, T., et al. 2005. Toxicogenomics in the pharma-ceutical industry: hollow promises or real benefit? *Mutat. Res.* 575: 102.

89. Tennant, R. W. 2002. The national center for toxicogenomics: using new technologies to inform mechanistic toxicology. *Environ. Health Perspect.* 110: 8.

90. Waters, M. D., and Fostel, J. M. 2004. Toxicogenomics and systems toxicology: aims and prospects. *Nat. Rev. Genet.* 5: 936.

91. Castle, A. L., Carver, M. P., and Mendrick, D. L. 2002. Toxicogenomics: a new revolu-tion in drug safety. *Drug Discov. Today* 7: 728.

92. Hayes, K. R., and Bradfield, C. A. 2005. Advances in toxicogenomics. *Chem. Res. Toxicol.* 18: 403.

93. Boverhof, D. R., and Zacharewski, T. R. 2006. Toxicogenomics in risk assessment: applications and needs. *Toxicol. Sci.* 89: 352.

94. Searfoss, G. H., Ryan, T. P., and Jolly, R. A. 2005. The role of transcriptome analysis in pre-clinical toxicology. *Curr. Mol. Med.* 5: 53.

95. Maggioli, J., Hoover, A., and Weng, L. 2006. Toxicogenomic analysis methods for pre-dictive toxicology. *J. Pharmacol. Toxicol. Methods* 53: 31.

96. Natsoulis, G., El Ghaoui, L., Lanckriet, G. R., Tolley, A. M., et al. 2005. Classification of a large microarray data set: algorithm comparison and analysis of drug signatures. *Genome Res.* 15: 724.

97. Cristianini, N., and Shawe-Taylor, J. 2000. *An introduction to support vector machines.* Cambridge, U.K.: Cambridge University Press.

98. Furey, T. S., Cristianini, N., Duffy, N., Bednarski, D. W., et al. 2000. Support vector machine classification and validation of cancer tissue samples using microarray expres-sion data. *Bioinformatics* 16: 906.

99. Kramer, J. A., Curtiss, S. W., Kolaja, K. L., Alden, C. L., et al. Acute molecular mark-ers of rodent hepatic carcinogenesis identified by transcription profiling. *Chem. Res. Toxicol.* 17: 463.

100. Ellinger-Ziegelbauer, H., Stuart, B., Wahle, B., Bomann, W., et al. Comparison of the expression profiles induced by genotoxic and nongenotoxic carcinogens in rat liver. *Mutat. Res.* 575: 61.

101. Food and Drug Administration, HHS. 1998. International Conference on Harmoni-zation: guidance on testing for carcinogenicity of pharmaceuticals. *Fed. Regist.* 63: 8983.

102. Liguori, M. J., Anderson, L. M., Bukofzer, S., McKim, J., et al. 2005. Microarray analy-sis in human hepatocytes suggests a mechanism for hepatotoxicity induced by trova-floxacin. *Hepatology* 41: 177.

103. Amin, K., Ip, C., Jimenez, L., Tyson, C., et al. 2005. In vitro detection of differential and cell-specific hepatobiliary toxicity induced by geldanamycin and 17-allylaminogel-danamycin using dog liver slices. *Toxicol. Sci.* 87: 442.

104. Waring, J. F., Ciurlionis, R., Jolly, R. A., Heindel, M., et al. 2001. Microarray analy-sis of hepatotoxins in vitro reveals a correlation between gene expression profiles and mechanisms of toxicity. *Toxicol. Lett.* 120: 359.

105. Sawada, H., Takami, K., and Asahi, S. 2005. A toxicogenomic approach to drug-induced phospholipidosis: analysis of its induction mechanism and establishment of a novel in vitro screening system. *Toxicol. Sci.* 83: 282.

106. Waring, J. F., Ciurlionis, R., Jolly, R. A., Heindel, M., et al. 2003. Isolated human hepa-tocytes in culture display markedly different gene expression patterns depending on attachment status. *Toxicol. In Vitro* 17: 693.
107. Li, A. P., Reith, M. K., Rasmussen, A., Gorski, J. C., et al. 1997. Primary human hepa-tocytes as a tool for the evaluation of structure-activity relationship in cytochrome P450 induction potential of xenobiotics: evaluation of rifampin, rifapentine and rifabu-tin. *Chem. Biol. Interact.* 107: 17.
108. Ulrich, R. G., Bacon, J. A., Cramer, C. T., Peng, G. W., et al. 1995. Cultured hepa-tocytes as investigational models for hepatic toxicity: practical applications in drug discovery and development. *Toxicol. Lett.* 82: 107.
109. de Longueville, F., Surry, D., Meneses-Lorente, G., Bertholet, V., et al. 2002. Gene expression profiling of drug metabolism and toxicology markers using a low-density DNA microarray. *Biochem. Pharmacol.* 64: 137.
110. Harries, H. M., Fletcher, S. T., Duggan, C. M., and Baker, V. A. 2001. The use of genomics technology to investigate gene expression changes in cultured human cells. *Toxicol. In Vitro* 15: 399.
111. Morgan, K. T., Ni, H., Brown, H. R., Yoon, L., et al. 2002. Application of cDNA micro-array technology to in vitro toxicology and the selection of genes for a real-time RT-PCR-based screen for oxidative stress in Hep-G2 cells. *Toxicol. Pathol.* 30: 435.
112. Aubrecht, J., and Caba, E. 2005. Gene expression profile analysis: an emerging approach to investigate mechanisms of genotoxicity. *Pharmacogenomics* 6: 419.
113. Snyder, R. D., and Green, J. W. 2001. A review of the genotoxicity of marketed phar-maceuticals. *Mutat. Res.* 488: 151.
114. Dickinson, D. A., Warnes, G. R., Quievryn, G., Messer, J., et al. 2004. Differentiation of DNA reactive and non-reactive genotoxic mechanisms using gene expression profile analysis. *Mutat. Res.* 549: 29.
115. Liszewski, K. 2006. Toxicogenomics enhances drug discovery. *Genetic Engineering News* 26: 40.
116. Hamadeh, H. K., Bushel, P. R., Jayadev, S., DiSorbo, O., et al. 2002. Prediction of com-pound signature using high density gene expression profiling. *Toxicol. Sci.* 67: 232.
117. Waring, J. F., Gum, R., Morfitt, D., Jolly, R. A., et al. 2002. Identifying toxic mecha-nisms using DNA microarrays: evidence that an experimental inhibitor of cell adhesion molecule expression signals through the aryl hydrocarbon nuclear receptor. *Toxicology* 181–182: 537.
118. Searfoss, G. H., Jordan, W. H., Calligaro, D. O., Galbreath, E. J., et al. 2003. Adip-sin, a biomarker of gastrointestinal toxicity mediated by a functional gamma-secretase inhibitor. *J. Biol. Chem.* 278: 46107.
119. Milano, J., McKay, J., Dagenais, C., Foster-Brown, L., et al. 2004. Modulation of notch processing by gamma-secretase inhibitors causes intestinal goblet cell metaplasia and induction of genes known to specify gut secretory lineage differentiation. *Toxicol. Sci.* 82: 341.
120. Zappa, F., Ward, T., Pedrinis, E., Butler, J., et al. 2003. NAD(P)H: quinone oxidoreduc-tase 1 expression in kidney podocytes. *J. Histochem. Cytochem.* 51: 297.
121. Su, A. I., Wiltshire, T., Batalov, S., Lapp, H., et al. 2004. A gene atlas of the mouse and human protein-encoding transcriptomes. *Proc. Natl. Acad. Sci. USA* 101: 6062.
122. Su, A. I., Cooke, M. P., Ching, K. A., Hakak, Y., et al. 2002. Large-scale analysis of the human and mouse transcriptomes. *Proc. Natl. Acad. Sci. USA* 99: 4465.
123. Honore, P., Kage, K., Mikusa, J., Watt, A. T., et al. 2002. Analgesic profile of intrathe-cal P2X(3) antisense oligonucleotide treatment in chronic inflammatory and neuro-pathic pain states in rats. *Pain* 99: 11.

124. Semizarov, D., Frost, L., Sarthy, A., Kroeger, P., et al. 2003. Specificity of short interfering RNA determined through gene expression signatures. *Proc. Natl. Acad. Sci. USA* 100: 6347.

125. Zambrowicz, B. P., Turner, C. A., and Sands, A. T. 2003. Predicting drug efficacy: knockouts model pipeline drugs of the pharmaceutical industry. *Curr. Opin. Pharmacol.* 3: 563.

126. Lu, P. Y., Xie, F., and Woodle, M. C. 2005. In vivo application of RNA interference: from functional genomics to therapeutics. *Adv. Genet.* 54: 117.

127. Li, L., Lin, X., Staver, M., Shoemaker, A., et al. 2005. Evaluating hypoxia-inducible factor-1alpha as a cancer therapeutic target via inducible RNA interference in vivo. *Cancer Res.* 65: 7249.

128. Corvi, R., Ahr, H. J., Albertini, S., Blakey, D. H., et al. 2006. Validation of toxicogenomics-based test systems: ECVAM-ICCVAM/NICEATM considerations for regulatory use. *Environ. Health Perspect.* 114: 420.

129. Venter, J. C., Adams, M. D., Myers, E. W., Li, P. W., et al. 2001. The sequence of the human genome. *Science* 291: 1304.

5 Gene Profiles and Cancer

Beverly A. Teicher
Genzyme Corporation

CONTENTS

5.1 INTRODUCTION

Cancer is a complex genetic disease involving multiple molecular abnormalities. Genetic events—including mutations, chromosomal gains, losses and rearrangements, along with epigenetic alterations—lead to significant transcriptional changes in cancer cells. Cancer diagnosis and prognosis is one of the most important emerging clinical applications of gene-expression profiling [1]. Gene-expression profiling of cancer tissues is expected to increase our understanding of cancer biology, as well as introduce new methods of cancer classification, diagnosis, and therapy [2]. Gene-expression profiling of normal tissues will most likely also facilitate our understanding of the etiology of specific organ diseases and augment the identification of targets for new therapeutics. Additionally, the genetics of cancer will provide illuminating insights into the interaction of environmental triggers with cancer genes, through gene profiling strategies. How this gene–environment interaction varies between individuals will certainly be a future inspiration for the development of individualized drugs, based on gene profiling and the extension of this knowledge to broader epigenetic influences.

5.2 GENE-EXPRESSION PROFILING IN CANCER

DNA microarrays are powerful tools, although not the only tools, for obtaining global views of tumor, normal tissue, and tumor tissue gene expression [3–7]. High-throughput RT-PCR (reverse transcriptase polymer chain reaction), serial analysis of gene expression (SAGE), subtractive hybridization, and high-resolution long oligonucleotide microarrays are other methods to define gene expression [8–13]. Currently, transcriptional profiling provides the most comprehensive approach for the analysis of somatic changes in pathways critical to the transformation of normal cells into malignant cells. Methodologies to analyze and interpret these data and software to facilitate the analyses have incorporated multicategory support vector machines, Bayesian networks, principal components analysis, self-organizing map-based clustering, and other comparative procedures [14–18]. Very recent work has elucidated the existence of small noncoding RNA species called microRNAs [19]. Lu et al. [19] used a bead-based flow cytometric miRNA (microRNA) expression profiling method to conduct a systematic expression analysis of 217 mammalian miRNAs in 334 samples and found that they were able to classify poorly differentiated tumors using miRNA expression profiles more accurately than using traditional mRNA (messenger RNA) profiling. Combined analysis of DNA microarray and drug-activity datasets has the potential of elucidating relationships between gene expression and drug sensitivity or resistance in the malignant cell as well as on-target and off-target effects of drugs [16, 20–22]. In addition, gene-expression profiling has been applied to understanding the responses of cells to mutagens and carcinogens such as ionizing radiation [23]. Anticancer immune responses can be enhanced by immune manipulation; however, the biological mechanism responsible for these immune responses remains largely unexplained [24]. Mocellin et al. [25] have applied high-throughput DNA array analysis toward understanding the molecular kinetics of immune response on a genome-wide scale, with a goal of elucidating cancer immunogenomics.

5.2.1 METASTASIS

The spread and growth of tumors at secondary sites, or metastasis, is extremely important clinically, since a majority of cancer mortality is associated with metastatic tumors [26]. In fact, metastasis has been described as the principal event leading to death in individuals with cancer. The genetic background from which a tumor arises can have significant effects on the ability of the tumor to metastasize and on gene-expression profiles. Ramaswamy et al. [27] identified a gene-expression signature that distinguished primary from metastatic adenocarcinomas and found that a subset of primary tumors resembled metastatic tumors. They found that solid tumors carrying the metastatic gene-expression signature were associated with metastasis and poor clinical outcome. Tumor hypoxia is an independent prognostic indicator of poor patient outcome. Hypoxia is a potent controller of gene expression, and identifying hypoxia-regulated genes is a means to investigate the molecular response to hypoxic stress [28]. Physiologically driven changes in hypoxic gene expression may provide insight into the poor outcomes for patients with hypoxic tumors.

5.2.2 DATA RELIABILITY AND APPLICATION TO THE CLINIC

Important to the process of incorporating gene-expression profiling into clinical practice is reproducibility of data from different laboratories. Dobbin et al. [29] assessed the comparability of data from four laboratories using Affymetrix Human Genome U133A arrays and found high within-laboratory and between-laboratory correlations on purified RNA samples, cell lines, and frozen tissues, thus adding confidence to the feasibility of combining data from independent laboratories for a single study. Although many studies of DNA microarray gene-expression profiling have been undertaken to predict cancer treatment outcome with a goal of tailoring treatment for patients based upon a knowledge of the transcript profile of their disease, this strategy has a way to go to be reliable. Michiels et al. [30] reanalyzed data from seven large studies that have attempted to predict the prognoses of cancer patients on the basis of DNA microarray analysis. They expanded the strategy by using more training sets and validation sets. The findings indicated that the list of genes identified as predictors of prognosis was highly unstable and that five of the seven studies did not classify patients better than chance. However, if accurate "prognostic signatures" including a relatively small number of genes can be developed, it is likely that these can become key decision-making tools in the clinic [31].

5.3 SPECIFIC CANCERS/TUMORS AND GENE PROFILING STRATEGIES

5.3.1 COLORECTAL CANCER

Genome-wide gene profiling using microarrays has identified many genes that are deregulated in colorectal cancer compared with normal tissue and allow classification of colorectal cancers [32–35]. Komori et al. [36] constructed gene-expression profiles of 50 colorectal cancers and 12 normal colorectal epithelia. There were 22 up-regulated genes and 32 down-regulated genes in the colorectal cancers compared with the normal epithelia. It is generally accepted that most colorectal carcinomas arise in existing adenomas. Adenomas can be divided into two groups: the protruded type and the flat type. Several studies have examined whether it is feasible to generate a molecular classification of adenomas and colorectal carcinomas by cDNA array gene expression [37]. Using cDNA arrays, Nosho et al. [37] analyzed the expression profiles of 550 cancer-related genes in 36 colorectal adenomas (18 flat and 18 protruded) and 14 early invasive carcinomas. By clustering analysis, they could distinguish the adenomas from the carcinomas and the flat from the protruded adenomas. Another study has examined gene-expression profiling in lymph node–positive and lymph node–negative colorectal cancer [38].

Liver resection is the primary therapy for patients with colorectal hepatic metastases. Liver metastasis is a major cause of colorectal-cancer-related mortality [39].

From microarray screening of 139 tumors, chemokine receptor CXCR4 gene expression in the primary tumor was found to be associated with recurrence and liver metastasis. Aschele et al. [40] examined the expression of deleted in colon cancer (DCC) protein in colorectal cancer metastases and primary tumors and examined the potential correlation of response to 5-fluorouracil with DCC expression. DCC was similarly expressed in the primary and metastatic tumors. DCC expression in the tumor was correlated with improved survival. Resistance to 5-fluorouracil is an obstacle in colorectal cancer therapy. Using GeneChip arrays, Schmidt et al. [41] found a set of 33 genes that distinguished sensitive and resistant cells; thus resistance to 5-fluorouracil involves a wide molecular repertoire. Li et al. [42] used the classic technique of subtractive hybridization followed by reverse northern dot blot analysis to examine differences in gene expression between primary colorectal cancers and liver metastases. BRCA1 was found to be down-regulated in the liver metastases. Other markers of metastasis for colorectal cancer include NY-ESO-1, LAGE-1, and MAGE-4 [43]. Array-based comparative genomic hybridization indicated that 30% ± 14% of the genomic fraction of the genome was altered in the metastases [44].

Wang et al. [45] used Affymetrix U133a GeneChip analysis to identify prognostic markers in Dukes' B patients. With samples from 31 patients whose tumors relapsed in less than 3 years and 43 patients who remained disease free at 3 years, a 23-gene signature was able to define the groups. Yu et al. [46] profiled the gene expression for 24 genes related to irinotecan activity in paired tumor and normal tissues from 52 patients with Dukes' C colorectal cancer using a real-time quantitative RT-PCR assay. Six of the 24 irinotecan pathway genes had lower expression in the tumor samples than in normal tissues, and 8 genes were much higher in the tumor tissues. Using hierarchical clustering, 3 gene clusters and 3 patients groups were observed with high similarity indices by RNA expressions in colorectal tumors.

5.3.2 Lung Cancer

Non-small cell lung cancer (NSCLC) is the most common cause of premature death from malignant disease in Western countries [47]. Gene-expression profiling is being used to enhance understanding of the molecular mechanisms underlying non-small cell lung cancer etiology, pathogenesis, and response to therapeutics [48, 49]. Multiple independent studies of gene/mRNA expression profiling in lung adenocarcinoma have shown reproducible alterations versus normal lung tissue [50]. Parmigiani et al. [51] performed a cross-study comparison of gene-expression studies for molecular classification of lung cancer using integrative correlations that quantify cross-study reproducibility without relying on direct assimilation of expression measurements across platforms. The analysis showed a large proportion of genes in which the patterns agreed across studies and identified genes that reproducibly predict outcomes of patients. Bioinformatics tools applied to lung cancer expression profiling data have identified prognostic genes that have been used to develop predictor models, but thus far, these models have not been incorporated into routine clinical use because of their inherent complexity and requirements for relatively large numbers of genes [52]. Gordon et al. [52] used ratios of gene expression to overcome these limitations and derived candidate ratio-based tests from analyses of 36 stage I lung adenocarcinoma samples. They went on to apply these tests to a set of 60 stage I lung adenocarcinomas and to stage II and III lung cancers. They conclude that expression ratios are highly accurate in predicting cancer recurrence.

Gene-expression profiles of resected tumors may predict treatment response and outcome. Forty-two genes from 23 tumor biopsies were associated with outcome [53]. The outcome training set class prediction accuracy rate was 87% in this study. Development of distant metastasis after primary tumor resection is the leading cause of death in early-stage non-small cell lung cancer [54]. Using quantitative RT-PCR, the expression of 56 receptor tyrosine kinases was determined in primary tumors from 70 patients [54]. There was a correlation between expression of several receptor tyrosine kinases and metastasis occurrence in these patients. Non-small cell lung cancer frequently expresses high levels of protein kinase C-alpha protein and mRNA; thus this cytosolic kinase may be a useful therapeutic target [55]. The expression of genes involved in nucleotide excision repair pathways was examined in 67 samples of resected lung cancers of stages IIB, IIIA, and IIIB treated with neo-adjuvant gemcitabine and cisplatin by RT-PCR [56]. A good correlation was found between mRNA expression of these genes and patient outcome. Patients having the lowest expression of these genes benefited the most from the neo-adjuvant chemotherapy.

Lymph node metastasis is not only an important factor in estimating the extent and the metastatic potential of an NSCLC, but also in prognosticating the patient outcome. Using a cDNA array, Takada et al. [57] analyzed the expression profiles of 1,289 genes in 92 non-small cell lung cancers (37 squamous cell carcinomas and 55 adenocarcinomas). Gene-expression profiles using oligonucleotide microarrays for 22 human non-small cell lung cancer cell lines identified 22 genes by cluster analysis that allowed classification of the cell lines into low- and high-invasive subgroups [58]. Using an optimized set of genes, it was possible to stratify the patients for lymph node metastasis and tumor stage. In a study of 47 patients, data correlations between gene-expression levels and tumor response to chemotherapy were found [59, 60].

Different vascular endothelial growth factor (VEGF) gene polymorphisms result in different VEGF gene responsiveness to various stimuli and different capacities for VEGF protein production [61]. Inherited VEGF sequence variations are strong determinants of molecular VEGF and VEGF-downstream phenotype of NSCLC. Genomic profiling was used to compare gene expression in early lung adenocarcinomas and non-neoplastic pulmonary tissue, with a focus on genes related to the control of apoptosis [62, 63]. The results of a study of 15 poorly differentiated lung adenocarcinomas and 5 normal pulmonary tissue samples found 24 apoptotic genes that were differentially expressed in the lung adenocarcinomas.

5.3.3 BREAST CANCER

Genomic expression profiling has greatly improved our ability to subclassify human breast cancers according to shared molecular characteristics and clinical behavior [64–68]. Gene-expression profiling holds the promise of more sensitive and specific indicators of prognosis to identify breast cancer patients at greatest risk for disease progression [69, 70]. Most invasive breast carcinomas are known to derive from precursor *in situ* lesions. One hypothesis is that major global expression abnormalities occur in the transition from normal to premalignant stages and further progression to invasive stages. Abba et al. [71] used serial analysis of gene expression (SAGE) to generate a comprehensive global gene-expression profile of the major changes occurring during breast cancer malignant evolution. In another study, Verlinden et al. [72] used microdissection and SAGE to identify genes expressed in ductal carcinoma *in situ* in the breast. Genomic analysis also has the potential to predict patient response to specific therapeutic agents, thus ensuring that patients derive the maximum benefit from the treatment they receive. Gene-expression profiles using the Affymetrix U133A chip system can be performed using both cryopreserved and fresh tissue biopsies [73].

Wang et al. [74] analyzed total RNA from 286 frozen tumor samples of lymph node–negative patients who had not received adjuvant systemic treatment and identified a 76-gene signature consisting of 60 genes for estrogen receptor–positive (ER+) patients and 16 genes for estrogen receptor–negative (ER–) patients. This gene signature provides a tool for identification of patients at high risk of distant recurrence [75]. Gene-expression patterns from an unselected group of 99 node-negative and node-positive breast cancer patients were found to be strongly associated with estrogen receptor (ER) status and moderately associated with grade, but were not associated with menopausal status, nodal status, or tumor size [76]. Microarray whole-genome gene-expression analysis, as well as quantitative analysis of mRNA expression of members of the estradiol metabolic and signaling pathways in tumors of postmenopausal breast cancer patients, point to a possible association between the development of a tumor with a particular expression profile and its capacity to synthesize estradiol, as measured by the expression of the transcripts for necessary pathway enzymes [77]. Using 6728-element cDNA microarrays, Gruvberger-Saal et al. [78] analyzed 48 well-characterized primary tumors from lymph node–negative breast cancer patients to determine a gene-profile-directed threshold for ER protein level to redefine the cutoff between ER-positive and ER-negative classes that may be more biologically relevant. To study gene expression of the malignant cells *in situ*, Verlinden et al. [72] used microdissection to separate ductal carcinoma *in situ* (DCIS) cells from the surrounding stroma, immunological infiltrates, and endothelial cells, and they applied microSAGE to the total mRNA to study the gene-expression profile. Reinholz et al. [79] examined the expression of genes related to transforming growth factor–beta for 14 normal breast, 5 noninvasive, 57 invasive, and 5 metastatic breast tumor tissues and found that mRNA levels of transforming growth factor–beta inducible early gene (TIEG) accurately discriminated between normal breast tissue and primary tumors with a sensitivity of 96% and 93%, respectively, with tumor expression being lower.

Predicting the metastatic potential of primary malignant tissues has direct bearing on the choice of therapy [80–83]. In patients with lymph node–negative breast carcinoma, overexpression of HER-2 significantly predicted higher risk of disease recurrence [84]. Docetaxel is one of the most active agents in breast cancer, but resistance or incomplete response is frequent [85]. Core biopsies from 24 patients were obtained before treatment with neoadjuvant docetaxel, and surgical specimens were obtained three months later. Differential expression of 92 genes correlated with docetaxel response, while gene-expression patterns that were very similar between the core biopsy and the surgical specimen correlated with resistance to the drug. In another study, Iwao-Koizumi et al. [86] examined 44 breast tumor tissues by biopsy before treatment with docetaxel. Gene-expression profiling of the biopsy samples was performed by RT-PCR. Using genes differentially expressed between responders and nonresponders, a diagnostic system based on a weighted-voting algorithm was constructed that provided over 80% accuracy in 26 previously unanalyzed samples. By comparing gene-expression changes prior to and shortly after the beginning of neo-adjuvant chemotherapy, it may be possible to identify patients with sensitive and resistant tumors and to help optimize the choice and schedule of the therapeutics [87].

Inflammatory breast cancer is a rare but particularly aggressive form of primary breast cancer [88]. Bieche et al. [88] used real-time quantitative PCR to quantify the mRNA expression of 538 selected genes and found 27 genes up-regulated in inflammatory breast cancer compared with non-inflammatory breast cancer, which may be useful for diagnostic or prognostic markers. In another study, Zhu et al. [89], using a combination of microdissection and microarray analysis, compared gene expression from cancer cells from five patients with ductal carcinoma in situ and separately from five patients with frankly invasive cancer and found gene-expression differences associated with microanatomical location of neoplastic cells, even in small developing tumor masses.

Gene-expression patterns between primary xenograft breast cancers and their matched metastases from lung and lymph nodes were examined [90]. A pattern of gene expression emerged that differentiated the primary from the metastatic disease. Tammen et al. [91] used peptide-based profiling of normal human mammary epithelial cells (HMEC) and the breast cancer cell line MCF-7 to examine the expression of up to 2300 peptides and found that about 8% of the peptides differed in abundance.

5.3.4 PROSTATE CANCER

The ability to distinguish between aggressive and nonaggressive prostate tumors has not changed despite vast improvements in the detection of prostate cancer [92, 93]. Global signature expression patterns obtained through multidimensional scaling appear to be promising in distinguishing between aggressive and nonaggressive forms of prostate cancer and in distinguishing prostate cancer from benign prostatic hyperplasia or normal prostate tissue. One of the major problems in the management of prostate cancer is the lack of reliable genetic markers predicting the clinical course of the disease. Glinsky et al. [94] analyzed gene-expression profiles in 21 prostate tumors to establish a recurrence predictor signature and then applied the recurrence predictor to 79 tumors. Overall, 88% of patients with recurrence of prostate cancer within 1 year after therapy were correctly classified into the poor prognosis group by the recurrence predictor algorithm. Bueno et al. [95] developed a ratio-based predictive model using a training set of 32 samples of prostate cancer and found that a three-ratio test using four genes was 90% accurate for distinguishing normal prostate and prostate cancer. Using "gene chips," Febbo and Sellers [96] were able to find gene-expression correlates of Gleason score, and a preliminary five-gene model was found that could separate cases into recurrent and nonrecurrent groups based on the expression patterns found in the primary tumors. Di Lorenzo et al. [97] retrospectively analyzed 126 prostate cancer samples from patients with clinically localized disease and radical prostatectomy (72) and from patients with metastatic androgen-independent disease (54). The results supported a role for bcl-2, COX-2, and angiogenesis in the development and progression of prostate cancer. Best et al. [98] measured and analyzed gene-expression profiles of 13 high- and moderate-grade human prostate tumors using cDNA microarrays. The expression of 136 genes was observed to differ significantly between normal prostate and tumors using one-sample t-test and Wilcoxon ranking. Hierarchical clustering of genes demonstrated a similar pattern across the tumors. However, permutation t-tests revealed 21 genes whose expression profiles segregated moderate- and high-grade tumors. Carcinoma-associated fibroblasts (CAF) promote tumor progression of pre-neoplastic epithelial cells. Transforming growth factor–beta 1 expression was higher in the carcinoma-associated stroma of prostate cancer tissue than the stroma of nonmalignant prostate tissue [99].

5.3.5 HEAD AND NECK CANCERS AND PANCREATIC CANCER

Head and neck squamous cell carcinoma is the fifth most common cancer worldwide. The advent of array-based technology and progress in the human genome initiative provide an opportunity to perform comprehensive molecular and genetic profiling of head and neck squamous cell carcinoma [100, 101]. Head and neck cancer is one of the most distressing human cancers, causing pain and affecting the basic survival functions of breathing and swallowing [102]. Mortality rates have not changed despite recent advances in radiotherapy and surgical treatment. For head and neck cancers, early detection of metastasis at lymph nodes close to the site of the primary tumor is pivotal for appropriate treatment. Roepman et al. [103], using 82 tumor samples, showed that DNA microarray gene-expression profiling can detect lymph node metastases for primary head and neck squamous cell carcinomas that arise in the oral cavity and oropharynx. A detailed genome mapping analysis of 213,636 expressed sequence tags (EST) derived from nontumor and tumor tissues of the oral cavity, larynx, pharynx, and thyroid was done [104]. Transcripts matching known human genes were identified; potential new splice variants were flagged and subjected to manual curation, pointing to 788 putatively new alternative splicing isoforms, the majority being insertion events. A microarray containing 17,840 complementary DNA clones was used to measure gene-expression changes associated with tumor progression in nine patients with squamous cell carcinoma of the oral cavity [105]. Samples were taken for analysis from the primary tumor, nodal metastasis, and "normal" mucosa from the patient's oral cavity. A distinct pattern of gene expression, with progressive up- or down-regulation of expression, is found during the progression from histological normal tissue to primary carcinoma and to nodal metastasis. Using 22 paired samples of head and neck squamous carcinoma and normal tissue from the same donors, Kuriakose et al. [106] applied a combinatorial approach to Affymetrix U95A chip microarray data to identify genes consistently up- or down-regulated and concluded that combinatorial analysis of microarray data is a promising technique for identifying differentially expressed genes with few false positives. Complementary DNA microarray analysis of nine human head and neck squamous cell carcinomas was used by Sok et al. [107] to produce a preliminary, comprehensive database of tumor-specific gene-expression profiles and an understanding of gene-expression changes implicated in carcinogenesis.

Invasive pancreatic cancer is thought to develop through a series of noninvasive duct lesions known as pancreatic intraepithelial neoplasia. Prasad et al. [108] used cDNA microarrays interrogating 15,000 transcripts to identify 49 genes that were differentially expressed in microdissected early pancreatic intraepithelial neoplasia compared with microdissected normal duct epithelium. These data suggest that pancreatic intraepithelial neoplasia development may involve Hedgehog-mediated conversion to a gastric epithelial differentiation program. Ohno et al. [109] established sublines of the human Panc-1 tumor cell line that were selectively metastatic to liver or to the peritoneum, and found that global gene expression in these three tumor cell lines could be used to differentiate the three lines and provide insight into human pancreatic cancer metastasis.

5.3.6 MELANOMA

Cutaneous malignant melanoma remains the leading cause of skin cancer death in industrialized countries [110, 111]. Using tissue microarrays, Alonso et al. [111] analyzed 165 malignant melanoma samples from 88 patients corresponding to distinct histological progression phases, radial, vertical, and metastases. The study identified expression profiles distinguishing specific melanoma progression stages. Malignant melanoma cells, even among cancer cells, are known for their genomic instability. Vaccination protocols based on autologous tumor material often require *in vitro* culturing of tumor cells [112]. Gene-expression changes in these cells during the culture period have been confirmed. Affymetrix and spotted oligonucleotide microarrays were used to assess global differential gene expression comparing normal human melanocytes with six melanoma cell strains from advanced lesions [113]. Validation at the protein level confirmed the overexpression in melanoma cells compared with normal melanocytes of several genes in the growth factor/receptor family that are known to confer growth advantage and metastatic capacity. Evidence for a role of the CXCR4 receptor and its ligand CXCL12 chemokine has been reported for melanomas [114]. The expression of CXCR4 in 31 of 71 primary cutaneous melanomas is well correlated with time to progression, indicating that CXCR4 expression could be an independent prognostic marker in primary cutaneous malignant melanomas.

5.3.7 OVARIAN CARCINOMA

Ovarian cancer is the fifth leading cause of cancer death for women in the United States and is the most lethal type of gynecologic cancer in the Western world [115, 116]. The high fatality rate is due in part because most ovarian cancer patients present with advanced disease. Profiles of gene transcription have begun to delineate the molecular basis of ovarian cancer, including distinctions between carcinomas of differing histology, tumor progression, and patient outcome [117–119]. Meinhold-Heerlein et al. [117] used oligonucleotide arrays to profile expressions of 12,500 genes in a series of 57 stage III serous ovarian adenocarcinomas from 52 patients, 8 with noninvasive borderline tumors and 44 with adenocarcinomas of varying grade. Well-differentiated invasive adenocarcinomas showed a strikingly similar profile to noninvasive borderline tumors, as compared with cancers of moderate or poor cellular differentiation, which were also highly similar. In a similar experiment, Santin et al. [118] analyzed the gene profiles of 10 primary ovarian serous papillary carcinoma cell lines, 2 established ovarian cancer cell lines, and 5 primary normal ovary epithelial cultures, and found the genes overexpressed in the established cell lines had little correlation with those overexpressed in primary ovarian cancer, highlighting the divergence of gene expression that occurs as a result of long-term culture.

Using oligonucleotide microarrays, Donninger et al. [115] profiled 37 advanced-stage papillary serous primary carcinomas and identified 1191 genes that were significantly differentially regulated between the ovarian cancer specimens and normal ovarian surface epithelia. TADG-12D (TMPRSS3) gene expression was examined by semiquantitative PCR in 50 ovarian tumors (41 adenocarcinomas, 3 noninvasive borderline tumors, and 6 adenomas) and 7 normal ovaries [120]. TADG-12D variant mRNA expression level in advanced clinical stage diseases was significantly higher than that in early-stage diseases in ovarian carcinomas, and mucinous adenocarcinomas expressed significantly higher levels than other tissue subtypes. The gene expressions of 20 ovarian carcinomas, 17 ovarian carcinomas metastatic to the omentum, and 50 normal ovaries were determined by Gene Logic Inc. using Affeymetrix GeneChip HU_95 arrays containing approximately 12,000 known genes [116]. Genes up-regulated in ovarian carcinoma tissue samples, compared with more than 300 other normal and diseased tissue samples, were identified. Lee et al. [121] used microarray analysis and found that the most differentially expressed genes in ovarian cancer linked to glucose/insulin metabolism, suggesting a possible molecular link between the glucose/insulin signaling pathway and the neoplasms of ovarian cancer. Lancaster et al. [122] used hierarchical clustering to identify patterns of gene expression that distinguished cancer from normal ovarian epithelium.

5.3.8 RENAL CELL CARCINOMA

Renal cell carcinoma is the most common cancer in adult kidney [123]. The accuracy of current diagnosis and prognosis of the disease and the effectiveness of the treatment for the disease are limited by the poor understanding of the disease at the molecular level. Renal cell carcinoma shows various clinical behaviors, and surgical modalities are currently the only effective therapy against this cancer [124, 125]. Ami et al. [124] examined gene-expression profiles from 10 renal cell carcinoma cell lines and 1 normal renal proximal tubular cell line, and found several genes up-regulated in the cell lines from metastatic lesions. Burczynski et al. [126] evaluated the association of gene-expression profiles in PBMCs (peripheral blood mononuclear cells) with clinical outcomes in patients with advanced renal cell cancer. Using supervised class prediction, they identified multivariate expression patterns in PBMCs capable of assigning favorable outcomes of time to death and time to progression in a test set of renal cancer patients, with overall performance accuracies of 72% and 85%, respectively.

Anaplasia is associated with therapy resistance and poor prognosis of Wilm's tumor [127]. The gene-expression profiling of 54 Wilm's tumors, 5 normal kidneys, and fetal kidney identified 97 clones representing 76 Unigenes or unclustered ESTs that clearly separate anaplastic Wilm's tumors from tumors with favorable histology. Renal medullary carcinoma is a rare kidney tumor with highly aggressive behavior [128]. This tumor occurs exclusively in young patients with the sickle cell trait or disease. Yang et al. [128] analyzed gene-expression profiles of 2 renal medullary carcinomas and 64 renal tumors and found significant differences in the gene-expression patterns, suggesting that the renal medullary carcinoma should not be treated as a conventional renal cell carcinoma but as a special malignancy.

5.3.9 GASTRIC CANCER AND BLADDER CANCER

Gastric cancer is one of the most frequently diagnosed malignancies in the world, especially in Korea and Japan, and is the second most common cause of cancer death [129, 130]. To understand the molecular mechanism associated with gastric carcinogenesis, Kim et al. [129] identified 40 genes up-regulated in gastric cancer using a novel 2K cDNA microarray based upon ESTs. Gastric tumor tissue and adjacent normal tissues from 21 patients allowed the identification of maspin as a gene frequently overexpressed in the tumors [131]. In a further study, 74 of 85 tumor tissues expressed maspin. From analysis of 90 gastric adenocarcinomas by DNA microarray gene expression, the chemokine (C-C motif) ligand 18 (CCL18) was identified as a gene of interest [132]. High CCL18 expression levels were associated with prolonged overall and disease-free patient survival in an array-based data set by univariate analysis. Gastrointestinal stromal tumors (GISTs) express KIT, a target for imatinib mesylate (*Gleevec*), and respond favorably to imatinib therapy. Went et al. [133] studied more than 3000 tumors from more than 120 different tumor categories by immunohistochemistry using seven commercially available anti-KIT antibodies. KIT was detected in 28 of 28 GISTs, 42 of 50 seminomas, and 34 of 52 adenoid-cystic carcinomas, but fewer melanomas, lung cancers, and few other tumors.

Bladder cancer is a common malignant disease characterized by frequent recurrences. At present there are no clinically useful markers available for identifying bladder cancer patients with a high risk of disease recurrence or progression [134, 135]. High-throughput microarrays are being used in expression-profiling analyses in bladder cancer studies dealing with both *in vitro* models and clinical specimens [136, 137]. Dyrskjot et al. [135] built a 32-gene molecular classifier using a cross-validation approach that was able to classify benign and muscle-invasive tumors with close correlation to pathological staging in an independent test set of 68 tumors. Pulmonary metastases frequently develop in patients with aggressive bladder cancer [138]. Nicholson et al. [138] derived three sublines of T24T human urothelial cancer cells that disseminate to lung and found 121 genes to be up-regulated in these sublines compared with the parental T24T cells. Four of these genes were confirmed in 23 primary human bladder-cancer lung metastases.

5.3.10 LYMPHOMA, LEUKEMIA, GLIOMAS, AND NEUROBLASTOMA

In the area of lymphoma and leukemia, gene-expression profiling is rapidly moving toward clinical application. Diffuse large B-cell lymphoma (DLBCL) is the most common type of non-Hodgkin's lymphoma and is an extremely heterogeneous disease [139]. DLBCL is now viewed as an amalgam of several different diseases that have distinct gene-expression profiles, oncogenic mechanisms, and clinical outcomes [140–149]. On the other hand, chronic lymphocytic leukemia (CLL) has a single gene-expression signature. In mantle cell lymphoma (MCL), gene expression associated with tumor proliferation is a useful predictor of survival. Based upon gene expressions from cDNA microarrays, DLBCL was classified into two gene-expression subgroups: (a) the germinal center B-cell-like (GCB) DLBCL subgroup that expressed genes characteristic of normal germinal center B cells and (b) the activated B-cell-like (ABC) DLBCL subgroup that expressed a subset of genes that are characteristic of plasma cells [145]. Wang et al. [148] used a statistical approach within a Bayesian framework to combine microarray data from three investigations of gene-expression profiling of B-cell chronic lymphocytic leukemia (CLL) and normal B cells (NBC) using three different microarray platforms, and they identified several genes that were consistently differentially expressed between the CLL and NBC samples. Cole et al. [149] used an RNA direct-labeling method to detect predictor genes for acute myeloid leukemia (AML) and acute lymphoblastic leukemia (ALL); thus direct labeling may be ideal for diagnostic applications that do not require target amplification. Gene-expression profiles of 10 children with acute lymphoblastic leukemia (ALL) were studied using cDNA arrays; the study found variability of gene expression for many genes and identified a few genes with consistent up- or down-regulation [143, 147].

Recent data suggest that morphologically indistinguishable glioblastomas have distinct classes of causal oncogene activation and that subclasses may be targeted by oncogene/signaling pathway-specific therapies [142]. Shai et al. [144] used Affymetrix high-density oligonucleotide arrays to identify the global gene-expression signatures associated with gliomas of different types and grades. The global transcriptional profiles of gliomas of different types and grades are distinct from each other and normal brain using a predictor constructed of 170 genes.

Patients with neuroblastoma are classified into risk groups according to the Children's Oncology Group risk stratification. Despite this stratification, the survival rate for patients with high-risk neuroblastoma remains <30% [150]. Wei et al. [150] performed gene-expression profiling using cDNA microarrays and artificial neural networks to develop an accurate predictor of survival for each individual patient with neuroblastoma. Nineteen predictor genes were able to partition Children's Oncology Group–stratified high-risk patients into two subgroups according to their survival status and thus predict prognosis independent of currently known risk factors.

5.4   EVOLUTION AND REVOLUTION IN TARGETING ANGIOGENESIS AS CANCER THERAPY

Based upon astute studies of tumor histochemistry, Algire and Chalkley understood that malignant tumors induced a vascular reaction leading to the growth of host capillaries into the tumor [151]. The central hypothesis of Algire and Chalkley was that vascular induction by solid tumors may be the major, and possibly the only, distinguishing factor leading to tumor growth beyond normal tissue control levels. By the late 1960s, Folkman and his colleagues had begun searching for a tumor angiogenesis factor [152–154]. Then, in his 1971 landmark report, Folkman proposed that antiangiogenesis might be a means of holding tumors in a nonvascularized dormant state [155]. In the 1980s, work done by several laboratories elucidated the pivotal role of vascular endothelial growth factor (VEGF) or vascular permeability factor (VPF) and its receptors in the regulation of normal and abnormal angiogenesis [156–158]. Dvorak proposed that an increase in microvascular permeability is associated with angiogenesis in tumors and wounds [157–159]. Thus, a major function of vascular permeability factor/vascular endothelial growth factor identified in the angiogenic process was the induction of plasma protein leakage, with the formation of a fibrin clot to serve as a substrate for tumor cell and endothelial cell growth [158, 159]. VEGF was also found to elicit a pronounced angiogenic response in a variety of *in vivo* models, including the chick chorioallantoic membrane [160].

The proliferation of blood vessels is crucial during varied physiological processes such as embryonic development, normal growth and differentiation, wound healing, and reproductive functions [161–163]. Many tumor cell lines secrete VEGF in culture [164]. *In situ* hybridization studies have demonstrated that VEGF mRNA is markedly higher in human tumors than in the corresponding normal tissue [165–172]. The data indicate that, in cancer, VEGF is a paracrine mediator secreted primarily by malignant cells and acting on nearby target endothelial cells [173, 174].

VEGF-mediated angiogenesis is a component of normal homeostasis, as in wound healing, and involves activation of endothelial cells from quiescence to migration, proliferation, and network formation and back to quiescence [175, 176]. VEGF is essential for normal embryonic vasculogenesis and angiogenesis. Partial inactivation of the VEGF gene through loss of a single allele results in embryonic lethality in mice [175, 177, 178]. Blocking VEGF activity, using VEGF inhibitors or anti-VEGF treatment, during the early neonatal period produces growth arrest, endothelial-cell apoptosis, and lethality [179]. VEGF-mediated angiogenesis is critical for normal ovarian and endometrial cycle function [180]. Treatment with VEGF inhibitors delays follicular development and suppresses luteal angiogenesis in rodents and primates [181, 182].

The central role of VEGF and the VEGF pathway in critical normal vasculature makes VEGF and VEGF-stimulating agents attractive as proangiogenic therapeutics for treatment of ischemic disease [183]. Some nonvascular cells express VEGF receptors −1 and −2, implying that there are additional biological properties for these growth factors [184]. Indeed, the VEGF family and its receptors determine the development and homeostasis of many organs, including the respiratory, skeletal, hematopoietic, nervous, renal, and reproductive systems, independent of their vascular role. Exercise produces a powerful angiogenic stimulus within the active muscle that leads to a functionally important increase in capillarity [185]. This effect has been associated with VEGF. Many proangiogenic therapeutic strategies rely on the premise that VEGF can induce new vasculature properly connected to existing vessels of an ill-perfused organ such as an ischemic myocardium [186]. Administration of VEGF induces massive accumulation of endothelial progenitor cells, and these cells coalesce into new vascular loops. With cessation of the VEGF stimulus, the new network undergoes hemodynamic remodeling to "normalize" the architecture.

Early withdrawal of VEGF results in the regression of most of the new vessels; therefore short stimulus with VEGF is unlikely to be effective therapeutically [186, 187].

5.5 CONCLUSION

Cancer is a group of complex diseases that involve a vast range of genetic lesions and malfunctions, as well as epigenetic influences. With such a complex genetic involvement, new gene-expression profiling systems and techniques offer unlimited future possibilities for identifying cancer therapy targets and diagnostic markers. Microarray technology, high-throughput RT-PCR, SAGE, and other molecular technologies provide the technical capacity to examine gene expression in cancer cells compared with normal, nontumor cells.

Gene-profiling techniques have identified deregulated/dysregulated genes in colorectal cancer and lung cancer, have helped subclassify breast cancers and provided prognostic markers, have helped distinguish aggressive from nonaggressive prostate cancer, and have assisted with the identification of progression stages in melanoma. Other tumor types will also benefit from genetic diagnosis and profiling (e.g., ovarian and bladder cancer).

REFERENCES

1. Yeatman, T. J. 2003. The future of clinical cancer management: one tumor, one chip. *Amer. Surgeon* 6669: 41.
2. Sultmann, H., and Poustka, A. 2004. Recent advances in transcription profiling of human cancer. *Curr. Opin. Molec. Therap.* 6: 593.
3. Rhodes, D. R., Yu, J., Shanker, K., Deshpande, N., Varambally, R., Ghosh, D., Barrette, T., Pandey, A., and Chinnaiyan, A. M. 2004. ONCOMINE: a cancer microarray database and integrated data-mining platform. *Neoplasia (NY)* 6: 1.
4. Bertucci, F., Viens, P., Tagett, R., Nguyen, C., Houlgatte, R., and Birnbaum, D. 2003. DNA arrays in clinical oncology: promises and challenges. *Lab. Invest.* 83: 305.
5. Son, C. G., Bilke, S., Davis, S., Greer, B. T., Wei, J. S., Whiteford, C. C., Chen, Q. R., Cenacchi, N., and Khnan, J. 2005. Database of mRNA gene expression profiles of multiple human organs. *Genome Res.*15: 443.
6. Centeno, B. A., Enkemann, S. A., Coppola, D., Huntsman, S., Bloom, G., and Yeatman, T. J. 2005. Classification of human tumors using gene expression profiles obtained after microarray analysis of fine-needle aspiration biopsy samples. *Cancer* 105: 101.
7. Wadlow, R., and Ramaswamy, S. 2005. DNA microarrays in clinical cancer research. *Curr. Molec. Med.* 5: 111.
8. Kato, K., Yamashita, R., Matoba, R., Monden, M., Noguchi, S., Takagi, T., and Nakai, K. 2005. Cancer gene expression database (CGED): a database for gene expression profiling with accompanying clinical information of human cancer tissues. *Nucl. Acids Res.* 33 (Database issue): D533.
9. Brennan, C., Zhang, Y., Leo, C., Feng, B., Cauwels, C., Aguirre, A. J., Kim, M., Protopopov, A., and Chin, L. 2004. High-resolution global profiling of genomic alterations with long oligonucleotide microarray. *Cancer Res.* 64: 4744.
10. Tuteja, R., and Tuteja, N. 2004. Serial analysis of gene expression (SAGE): application in cancer research. *Med. Sci. Monitor* 10: RA132.
11. Porter, D., and Polyak, K. 2003. Cancer target discovery using SAGE. *Exp. Opin. Therap. Targets* 7: 759.

12. Hermeking, H. 2003. Serial analysis of gene expression and cancer. *Curr. Opin. Oncol.* 15: 44.
13. Alexandridis, R., Lin, S., and Irwin, M. 2004. Class discovery and classification of tumor samples using mixture modeling of gene expression data—a unified approach. *Bioinformatics* 20: 2545.
14. Statnikov, A., Aliferis, C. F., Tsamardinos, I., Hardin, D., and Levy, S. 2005. A comprehensive evaluation of multicategory classification methods for microarray gene expression cancer diagnosis. *Bioinformatics* 21: 631.
15. Covell, D. G., Wallqvist, A., Rabow, A. A., and Thanki, N. 2003. Molecular classification of cancer: unsupervised self-organizing map analysis of gene expression microarray data. *Molec. Cancer. Therap.* 2: 317.
16. Chang, J. H., Hwang, K. B., Oh, S. J., and Zhang, B. T. 2005. Bayesian network learning with feature abstraction for gene-drug dependency analysis. *J. Bioinformatics Comp. Biol.* 3: 61.
17. Takahashi, H., Kobayashi, T., and Honda, H. 2005. Construction of robust prognostic predictors by using projective adaptive resonance theory as a gene filtering method. *Bioinformatics* 21: 179.
18. Iacobuzio-Donahue, C. A., Ashfaq, R., Maitra, A., Adsay, N. V., Shen-Ong, G. L., Berg, K., Hollingsworth, M. A., Cameron, J. L., Yeo, C. J., Kern, S. E., Goggins, M., and Hruban, R. H. 2003. Highly expressed genes in pancreatic ductal adenocarcinomas: a comprehensive characterization and comparison of the transcription profiles obtained from three major technologies. *Cancer Res.* 63: 8614.
19. Lu, J., Getz, G., Miska, E. A., Alvarez-Saavedra, E., Lamb, J., Peck, D., Sweet-Cordero, A., Ebert, B. L., Mak, R. H., Ferrando, A. A., Downing, J. R., Jacks, T., Horvitz, H. R., and Golub, T.,R. 2005. MicroRNA expression profiles classify human cancers *Nature* 435: 834.
20. Clarke, P. A., te Poele, R., and Workman, P. 2004. Gene expression microarray technologies in the development of new therapeutic agents. *Eur. J. Cancer* 40: 2560.
21. Tanaka, T., Tanimoto, K., Otani, K., Satoh, K., Ohtaki, M., Yoshida, K., Toge, T., Yahata, H., Tanaka, S., Chayama, K., Okazaki, Y., Hayashizaki, Y., Hiyama, K., and Nishiyama, M. 2004. Concise prediction models of anticancer efficacy of 8 drugs using expression data from 12 selected genes. *Int. J. Cancer* 111: 617.
22. Dracopoli, N. C. 2005. Development of oncology drug response markers using transcriptional profiling. *Curr. Molec. Med.* 5: 103–110.
23. Snyder, A. R., and Morgan, W. F. 2004. Gene expression profiling after irradiation: clues to understanding acute and persistent responses? *Cancer Met. Rev.* 23: 259.
24. Wang, E., Panelli, M. C., Monsurro, V., and Marincola, F. M. 2004. Gene expression profiling of anticancer immune responses. *Curr. Opin. Molec. Therap.* 6: 288.
25. Mocellin, S., Wang, E., Panelli, M., Pilati, P., and Marincola, F. M. 2004. DNA array-based gene profiling in tumor immunology. *Clin. Cancer Res.* 10: 4597.
26. Hunter, K. W. 2004. Host genetics and tumor metastasis. *Brit. J. Cancer* 90: 752.
27. Ramaswamy, S., Ross, K. N., Lander, E. S., and Golub, T. R. 2003. A molecular signature of metastasis in primary solid tumors. *Nature Genetics* 33: 49.
28. Denko, N. C., Fontana, L. A., Hudson, K. M., Sutphin, P. D., Raychaudhuri, S., Altman, R., and Giaccia, A. J. 2003. Investigating hypoxic tumor physiology through gene expression patterns. *Oncogene* 22: 5907.
29. Dobbin, K. K., Beer, D. G., Meyerson, M., Yeatman, T. J., Gerald, W. L., Jacobson, J. W., et al. 2005. Interlaboratory comparability study of cancer gene expression analysis using oligonucleotide microarrays. *Clin. Cancer Res.* 11: 565.
30. Michiels, S., Koscielny, S., and Hill, C. 2005. Prediction of cancer outcome with microarrays: a multiple random validation strategy. *Lancet* 365: 488.

31. Bucca, G., Carruba, G., Saetta, A., Muti, P., Castagnetta, L., and Smith C. P. 2004. Gene expression profiling of human cancers. *Ann. N.Y. Acad. Sci.* 1028: 28.
32. Shih, W., Chetty, R., and Tsao, M. S. 2005. Expression profiling by microarrays in colorectal cancer. *Oncology Reps.* 13: 517.
33. Bustin, S. A., and Dorudi, S. 2004. Gene expression profiling for molecular staging and prognosis prediction in colorectal cancer. *Expert Rev. Molecular Diagnostics* 4: 599.
34. Mariadason, J. M., Arango, D., and Augenlicht, L. H. 2004. Customizing chemotherapy for colon cancer: the potential of gene expression profiling. *Drug Resistance Updates* 7: 209.
35. Tsunoda, T., Koh, Y., Koizumi, F., Tsukiyama, S., Ueda, H., Taguchi, F., Yamaue, H., Saijo, N., and Nishio, K. 2003. Differential gene expression profiles and identification of the genes relevant to clinicopathologic factors in colorectal cancer selected by cDNA array in combination with principal component analysis. *Int. J. Oncol.* 23: 49.
36. Komori, T., Takemasa, I., Higuchi, H., Yamasaki, M., Ikeda, M., Yamamoto, H., Ohue, M., Nakamori, S., Sekimoto, M., Matsubara, K., and Monden, M. 2004. Identification of differentially expressed genes involved in colorectal carcinogenesis using a cDNA microarray. *J. Exp. Clin. Cancer Res.* 23: 521.
37. Nosho, K., Yamamoto, H., Adachi, Y., Endo, T., Hinoda, Y., and Imai, K. 2005. Gene expression profiling of colorectal adenomas and early invasive carcinomas by cDNA array analysis. *Brit. J. Cancer* 92: 1193.
38. Kwon, H. C., Kim, S. H., Roh, M. S., Kim, J. S., Lee, H. S., Choi, H. J., Jeong, J. S., Kim, H. J., and Hwang, T. H. 2004. Gene expression profiling in lymph node-positive and lymph node-negative colorectal cancer. *Diseases Colon Rectum* 47: 141.
39. Kim, J., Takeuchi, H., Lam, S. T., Turner, R. R., Wang, H. J., Kuo, C., Foshag, L., Bilchik, A. J., and Hoon, D. S. 2005. Chemokine receptor CXCR4 expression in colorectal cancer patients increases the risk for recurrences and for poor survival. *J. Clin. Oncol.* 23: 2744.
40. Aschele, C., Debernardis, D., Lonardi, S., Bandelloni, R., Casazza, S., Monfardini, S., and Gallo, L. 2004. Deleted in colon cancer protein expression in colorectal cancer metastases: a major predictor of survival in patients with unresectable metastatic disease receiving palliative fluorouracil-based chemotherapy. *J. Clin. Oncol.* 22: 3758.
41. Schmidt, W. M., Kalipciyan, M., Dornstauder, E., Rizovski, B., Steger, G. G., Sedivy, R., Mueller, M. W., and Mader, R. M. 2004. Dissecting progressive stages of 5-fluorouracil resistance in vitro using RNA expression profiling. *Int. J. Cancer* 112: 200.
42. Li, M., Yuan, Y. H., Han, Y., Liu, Y. X., Yan, L., Wang, Y., and Gu, J. 2005. Expression profile of cancer testis genes in 121 human colorectal cancer tissue and adjacent normal tissue. *Clin. Cancer Res.* 11: 1809.
43. Li, S. R., Dorudi, S., and Bustin, S. A. 2003. Identification of differentially expressed genes associated with colorectal cancer liver metastasis. *Eur. Surg. Res.* 35: 327.
44. Mehta, K. R., Nakao, K., Zuraek, M. B., Ruan, D. T., Bergsland, E. K., Venook, A. P., et al. 2005. Fractional genomic alteration detected by array-based comparative genomic hybridization independently predicts survival after hepatic resection for metastatic colorectal cancer. *Clin. Cancer Res.* 11: 1791.
45. Wang, Y., Atkoe, T., Zhang, Y., Mutch, M. G., Talantov, D., Jiang, J., McLeod, H. L., and Atkins, D. 2004. Gene expression profiles and molecular markers to predict recurrence of Dukes' B colon cancer. *J. Clin. Oncol.* 22: 1564.
46. Yu, J., Shannon, W. D., Watson, M. A., and McLeod, H. L. 2005. Gene expression profiling of the irinotecan pathway in colorectal cancer. *Clin. Cancer Res.* 11: 2053.
47. Petty, R. D., Nicolson, M. C., Kerr, K. M., Collie-Duguid, E., and Murray, G. I. 2004. Gene expression profiling in non-small cell lung cancer: from molecular mechanisms to clinical application. *Clin. Cancer Res.* 10: 3237.

48. Muller-Hagen, G., Beinert, T., and Sommer, A. 2004. Aspects of lung cancer gene expression profiling. *Curr. Opin. Drug Discovery Develop.* 7: 290.

49. Meyerson, M., Franklin, W. A., and Kelley, M. J. 2004. Molecular classification and molecular genetics of human lung cancers. *Semin. Oncol.* 31: 4.

50. Meyerson, M., and Carbone, D. 2005. Genomic and proteomic profiling of lung cancers: lung cancer classification in the age of targeted therapy. *J. Clin. Oncol.* 23: 3219.

51. Parmigiani, G., Garrett-Mayrer, E. S., Anbazhagan, R., and Gabrielson, E. 2004. A cross-study comparison of gene expression studies for the molecular classification of lung cancer. *Clin. Cancer Res.* 10: 2922.

52. Gordon, G. J., Richards, W. G., Sugarbaker, D. J., Jaklitsch, M. T., and Bueno, R. 2003. A prognostic test for adenocarcinoma of the lung from gene expression profiling data. *Cancer Epidem. Biomarkers Prevent.* 12: 905.

53. Borczuk, R., Shah, L., Pearson, G. D., Walter, K. L., Wang, L., Austin, J. H., Friedman, R. A., and Powell, C. A. 2004. Molecular signatures in biopsy specimens of lung cancer. *Am. J. Resp. Crit. Care Med.* 170: 167.

54. Muller-Tidow, C., Diederichs, S., Bulk, E., Pohle, T., Steffen, B., Schwable, J., et al. 2005. Identification of metastasis-associated receptor tyrosine kinases in non-small cell lung cancer. *Cancer Res.* 65: 1778.

55. Lahn, M., Su, C., Li, S., Chedid, M., Hanna, K. R., Graff, J. R., et al. 2004. Expression levels of protein kinase C-alpha in non-small cell lung cancer. *Clin. Lung Cancer* 6: 184.

56. Rosell, R., Felip, E., Taron, M., Majo, J., Mendez, P., Sanchez-Ronco, M., Queralt, C., Sanchez, J. J., and Maestre, J. 2004. Gene expression as a predictive marker of outcome in stage IIB-IIIA-IIIB non-small cell lung cancer after induction gemcitabine-based chemotherapy followed by resectional surgery. *Clin Cancer Res.* 10: 4215s.

57. Takada, M., Tada, M., Tamoto, E., Kawakami, A., Murrakawa, K., Shindoh, G., et al. 2004. Prediction of lymph node metastasis by analysis of gene expression profiles in non-small cell lung cancer. *J. Surg. Res.* 122: 61.

58. Lader, A. S., Ramoni, M. F., Zetter, B. R., Kohane, I. S., and Kwiatkowski, D. J. 2004. Identification of a transcriptional profile associated with in vitro invasion in non-small cell lung cancer cell lines. *Cancer Biol. Therap.* 3: 624.

59. Oshita, F., Ikehara, M., Sekiyama, Y., Hamanaka, N., Saito, H., Yamada, K., Noda, K., Kameda, Y., and Miyagi, Y. 2004. Genomic-wide cDNA microarray screening to correlate gene expression profile with chemoresistance in patients with advanced lung cancer. *J. Exp. Therap. Oncol.* 4: 155.

60. Ikehara, M., Oshita, F., Sekiyama, A., Hamamaka, N., Saito, H., Yamada, K., Noda, K., Kameda, Y., and Miyagi, Y. 2004. Genome-wide cDNA microarray screening to correlate gene expression profile with survival in patients with advanced lung cancer. *Oncol. Reps.* 11: 1041.

61. Koukourakis, M. I., Papazoglou, D., Giatromanolaki, A., Bougioukas, G., Maltezos, E., and Siviridis, E. 2004. VEGF gene sequence variation defines VEGF gene expression status and angiogenic activity in non-small cell lung cancer. *Lung Cancer* 46: 293.

62. Singhal, S., Amin, K. M., Kruklitis, R., Marshall, M. B., Kucharczuk, J. C., DeLong, P., Litzky, L. A., Kaiser, L. R., and Albelda, S. M. 2003. Differentially expressed apoptotic genes in early stage lung adenocarcinoma predicted by expression profiling. *Cancer Biol. Therap.* 2: 566.

63. Sasaki, H., Moriyama, S., Mizuno, K., Yukiue, H., Yano, M., Fukai, I., Yamakawa, Y., and Fujii, Y. 2003. SAGE mRNA expression in advanced-stage lung cancers. *Eur. J. Surg. Oncol.* 29: 900.

64. Wilson, C. A., and Dering, J. 2004. Recent translational research: microarray expression profiling of breast cancer—beyond classification and prognostic markers? *Breast Cancer Res.* 6: 192.

65. Cleator, S., and Ashworth, A. 2004. Molecular profiling of breast cancer: clinical implications. *Brit. J. Cancer* 90: 1120.
66. Wang, Z. C., Lin, M., Wei, L. J., Li, C., Miron, A., Lodeiro, G., Harris, L., Ramaswamy, S., Tanenbaum, D. M., Meyerson, M., Iglehart, J.,D., and Richardson, A. 2004. Loss of heterozygosity and its correlation with expression profiles in subclasses of invasive breast cancers. *Cancer Res.* 64: 64.
67. Ma, X. J., Salunga, R., Tuggle, J. T., Gaudet, J., Enright, E., McQuary, P., et al. 2003. Gene expression profiles of human breast cancer progression. *Proc. Natl. Acad. Sci. USA* 100: 5974.
68. Bertucci, F., Viens, P., Hingamp, P., Nasser, V., Houlgatte, R., and Birnbaum, D. 2003. Breast cancer revisited using DNA array-based gene expression profiling. *Int. J. Cancer* 103: 565.
69. Gradishar, W. J. 2005. The future of breast cancer: the role of prognostic factors. *Breast Cancer Res. Treat.* 89 (Suppl 1): S17.
70. Ellis, M. J. 2003. Breast cancer gene expression analysis—the case for dynamic profiling. *Adv. Exp. Med. Biol.* 532: 223.
71. Abba, M. C., Drake, J. A., Hawkins, K. A., Hu, Y., Sun, H., Notcovich, C., Gaddis, S., Sahin, A., Baggerly, K., and Aldaz, C. M. 2004. Transcriptome changes in human breast cancer progression as determined by serial analysis of gene expression (SAGE). *Breast Cancer Res.* 6: R499.
72. Verlinden, I., Janssens, J., Raus, J., and Michiels, L. 2004. Microdissection and SAGE as a combined tool to reveal gene expression in ductal carcinoma in situ of the breast. *Molec. Carcinogenesis* 41: 197.
73. Drubin, D., Smith, J. S., Liu, W., Zhao, W., Chase, G. A., and Clawson, G. A. 2005. Comparison of cryopreservation and standard needle biopsy for gene expression profiling of human breast cancer specimens. *Breast Cancer Res. Treat.* 90: 93.
74. Wang, Y., Klijn, J. G., Zhang, Y., Sieuwerts, A.,M., Look, M. P., Yang, F., Talantov, D., Timmermans, M., Meijer-van Gelder, M. E., Yu, J., Jatkoe, T., Berns, E. M., Atkins, D., and Foekens, J.,A. 2005. Gene-expression profiles to predict distant metastasis of lymph-node-negative primary breast cancer. *Lancet* 365: 671.
75. Huang, E., West, M., and Nevins, J. R. 2003. Gene expression profiling for prediction of clinical characteristics of breast cancer. *Recent Prog. Hormone Res.* 58: 55.
76. Sotiriou, C., Neo, S. Y., McShane, L. M., Korn, E. L., Long, P. M., Jazaeri, A., Martiat, P., Fox, S. B., Harris, A. L., and Liu, E. T. 2003. Breast cancer classification and prognosis based on gene expression profiles from a population-based study. *Proc. Natl. Acad. Sci. USA* 100: 10393.
77. Kristensen, V. N., Sorlie, T., Geisler, J., Langerod, A., Yoshimura, N., Karesen, R., Harada, N., Lonning, P. E., and Borresen-Dale, A. L. 2005. Gene expression profiling of breast cancer in relation to estrogen receptor status and estrogen-metabolizing enzymes: clinical implications. *Clin. Cancer Res.* 11: 878s.
78. Gruvberger-Saal, S. K., Eden, P., Ringner, M., Baldetorp, B., Chebil, G., Borg, A., Ferno, M., Peterson, C., and Meltzer, P. S. 2004. Predicting continuous values of prognostic markers in breast cancer from microarray gene expression profiles. *Molec. Cancer Therap.* 3: 161.
79. Reinholz, M. M., An, M. W., Johnsen, S. A., Subramaniam, M., Suman, V. J., Ingle, J. N., Roche, P. C., and Spelsberg, T. C. 2004. Differential gene expression of TGF-beta inducible early gene (TIEG), Smad7, Smad2 and Bard1 in normal and malignant breast tissue. *Breast Cancer Res. Treat.* 86: 75.
80. Ein-Dor, L., Kela, I., Getz, G., Givol, D., and Domany, E. 2005. Outcome signature genes in breast cancer: is there a unique set? *Bioinformatics* 21: 171.

81. Miller, D. V., Leontovich, A. A., Lingle, W. L., Suman, V. J., Mertens, M.,L., Lillie, J., Ingalls, K. A., Perez, E. A., Ingle, J. N., Couch, F. J., and Visscher, D. W. 2004. Utilizing Nottingham Prognostic Index in microarray gene expression profiling of breast carcinomas. *Modern Pathology* 17: 756.

82. Weigelt, B., Verduijn, P., Bosma, A. J., Rutgers, E. J., Peterse, H. L., and Van't Veer, L. J. 2004. Detection of metastases in sentinel lymph nodes of breast cancer patients by multiple mRNA markers. *Brit. J. Cancer* 90: 1531.

83. Weigelt, B., Glas, A. M., Wessels, L. F., Witteveen, A. T., Peterse, J. L., and van't Veer, L. J. 2003. Gene expression profiles of primary breast tumors maintained in distant metastases. *Proc. Natl. Acad. Sci. USA* 100: 15901.

84. Sun, J. M., Han, W., Im, S. A., Kim, T. Y., Park, I. A., Noh, D. Y., Heo, D. S., Bang, Y. J., Choe, K. J., and Kim, N. K. 2004. A combination of HER-2 status and the St. Gallen classification provides useful information on prognosis in lymph-node-negative breast carcinoma. *Cancer* 101: 2516.

85. Chang, J. C., Wooten, E. C., Tsimelzon, A., Hilsenback, S. G., Gutierrez, M. C., Tham, Y. L., et al. 2005. Patterns of resistance and incomplete response to docetaxel by gene expression profiling in breast cancer patients. *J. Clin. Oncol.* 23: 1169.

86. Iwao-Koizumi, K., Matoba, R., Ueno, N., Kim, S. J., Ando, A., Miyoshi, Y., Maeda, E., Noguchi, S., and Kato, K. 2005. Prediction of docetaxel response in human breast cancer by gene expression profiling. *J. Clin. Oncol.* 23: 422.

87. Modlich, O., Prisack, H. B., Munnes, M., Audretsch, W., and Bojar, H. 2004. Immediate gene expression changes after the first course of neoadjuvant chemotherapy in patients with primary breast cancer disease. *Clin. Cancer Res.* 10: 6418.

88. Bieche, I., Lerebours, F., Tozlu, S., Espie, M., Marty, M., and Lidereau, R. 2004. Molecular profiling of inflammatory breast cancer: identification of a poor-prognosis gene expression signature. *Clin. Cancer Res.* 10: 6789.

89. Zhu, G., Reynolds, L., Crnogorac-Jurcevic, T., Gillett, C. E., Dublin, E. A., Marshall, J. F., Barnes, D., D'Arrigo, C., Van Trappen, P. O., Lemoine, N. R., and Hart, I. R. 2003. Combination of microdissection and microarray analysis to identify gene expression changes between differentially located tumor cells in breast cancer. *Oncogene* 22: 3742.

90. Montel, V., Huang, T. Y., Mose, E., Pestonjamasp, K., and Tarin, D. 2005. Expression profiling of primary tumors and matched lymphatic and lung metastases in a xenogeneic breast cancer model. *Am. J. Pathol.* 166: 1565.

91. Tammen, H., Kreipe, H., Hess, R., Kellmann, M., Lehmann, U., Pich, A., Lamping, N., Schulz-Knappe, P., Zucht, H. D., and Lilischkis, R. 2003. Expression profiling of breast cancer cells by differential peptide display. *Breast Cancer Res. Treat.* 79: 83.

92. Huppi, K., and Chandramouli, G. V. 2004. Molecular profiling of prostate cancer. *Curr. Urology Rep.* 5: 45.

93. Gerald, W. L. 2003. Genome-wide gene expression analysis of prostate carcinoma. *Semin. Oncol.* 30: 635.

94. Glinsky, G. V., Glinskii, A. B., Stephenson, A. J., Hoffman, R. M., and Gerald, W. L. 2004. Gene expression profiling predicts clinical outcome of prostate cancer. *J. Clin. Invest.* 113: 913.

95. Bueno, R., Loughlin, K. R., Powell, M. H., and Gordon, G. J. 2004. A diagnostic test for prostate cancer from gene expression profiling data. *J. Urology* 171: 903.

96. Febbo, P. G., and Sellers, W. R. 2003. Use of expression analysis to predict outcome after radical prostatectomy. *J. Urology* 170: S11.

97. Di Lorenzo, G., De Placido, S., Autorino, R., De Laurentiis, M., Mignogna, C., D'Armiento, M., Tortora, G., De Rosa, G., D'Armiento, M., De Sio, M., Bianco, A. R., and D'Armiento, F. P. 2005. Expression of biomarkers modulating prostate cancer progression: implications in the treatment of the disease. *Prostate Cancer Prostatic Diseases* 8: 54.

98. Best, C. J., Leviva, I. M., Chuaqui, R. F., Gillespie, J. W., Duray, P. H., Murgai, M., et al. 2003. Molecular differentiation of high- and moderate-grade human prostate cancer by cDNA microarray analysis. *Diagnostic Molec. Path.* 12: 63.

99. San Francisco, I. F., DeWolf, W. C., Peehl, D. M., and Olumi, A. F. 2004. Expression of transforming growth factor beta-1 and growth in soft agar differentiate prostate carcinoma-associated fibroblasts from normal prostate fibroblasts. *Int. J. Cancer* 112: 213.

100. Sotiriou, C., Lothaire, P., Dequanter, D., Cardoso, F., and Awada, A. 2004. Molecular profiling of head and neck tumors. *Curr. Opin. Oncol.* 16: 211.

101. Lemaire, F., Millon, R., Young, J., Cromer, A., Wasylyk, C., Schultz, I., Muller, D., Marchal, P., Zhao, C., Melle, D., Bracco, L., Abecassis, J., and Wasylyk, B. 2003. Differential expression profiling of head and neck squamous cell carcinoma (HNSCC). *Brit. J. Cancer* 89: 1940.

102. Chin, D., Boyle, G. M., Williams, R. M., Ferguson, K., Pandeya, N., Pedley, J., Campbell, C. M., Theile, D. R., Parsons, P. G., and Coman W. B. 2005. Novel markers for poor prognosis in head and neck cancer. *Int. J. Cancer* 113: 789.

103. Roepman, P., Wessels, L. F., Kettelarij, N., Kemmeren, P., Miles, A. J., Lijnzaad, P., et al. 2005. An expression profile for diagnosis of lymph node metastases from primary head and neck squamous cell carcinomas. *Nature Genetics* 37: 182.

104. Reis, E. M., Ojopi, E. P., Alberto, F. L., Rahal, P., Tsukumo, F., Mancini, U. M., et al. 2005. Large-scale transcriptome analyses reveal new genetic marker candidates of head, neck and thyroid cancer. *Cancer Res.* 65: 1693.

105. Belbin, T. J., Singh, B., Smith, R. V., Socci, N. D., Wreesmann, V. B., Sanchez-Carbayo, M., et al. 2005. Molecular profiling of tumor progression in head and neck cancer. *Arch. Otolaryngology H&N Surg.* 13: 10.

106. Kuriakose, M. A., Chen, W. T., He, Z. M., Sikora, A. G., Zhang, P., Zhang, Z. Y., et al. 2004. Selection and validation of differentially expressed genes in head and neck cancer. *Cell Molec. Life Sci.* 61: 1372.

107. Sok, J. C., Kuriakose, M. A., Mahajan, V. B., Pearlman, A. N., DeLacure, M. D., and Chen, F. A. 2003. Tissue-specific gene expression of head and neck squamous cell carcinoma in vivo by complementary DNA microarray analysis. *Arch. Otolaryngol. H&N Surg.* 129: 760.

108. Prasad, N. B., Biankin, A. V., Fukushima, N., Maitra, A., Dhara, S., Elkahloun, A. G., Hruban, R. H., Goggins, M., and Leach, S. D. 2005. Gene expression profiles in pancreatic intraepithelial neoplasia reflect the effects of hedgehog signaling on pancreatic ductal epithelial cells. *Cancer Res.* 65: 1619.

109. Ohno, K., Hata, F., Nishimori, H., Yasoshima, T., Yanai, Y., Sogahata, K., et al. 2003. Metastatic-associated biological properties and differential gene expression profiles in established highly liver and peritoneal metastatic cell lines of human pancreatic cancer. *J. Exp. Clin. Cancer Res.* 22: 6230.

110. Carr, K. M., Bittner, M., and Trent, J. M. 2003. Gene-expression profiling in human cutaneous melanoma. *Oncogene* 22: 3076.

111. Alonso, S. R., Ortiz, P., Pollan, M., Perez-Gomez, B., Sanchez, L., Acuna, M. J., et al. 2004. Progression in cutaneous malignant melanoma is associated with distinct expression profiles: a tissue microarray-based study. *Am. J. Pathol.* 164: 193.

112. Vogl, A., Sartorius, U., Vogt, T., Roesch, A., Landthaler, M., Stolz, W., and Becker, B. 2005. Gene expression profile changes between melanoma metastases and their daughter cell lines: implication for vaccination protocols. *J. Invest. Dermatol.* 124: 401.

113. Hoek, K., Rimm, D. L., Williams, K. R., Zhao, H., Ariyan, S., Lin, A., et al. 2004. Expression profiling reveals novel pathways in the transformation of melanocytes to melanomas. *Cancer Res.* 64: 5270.

114. Scala, S., Ottaiano, A., Ascierto, P. A., Cavalli, M., Simeone, E., Giuliano, P., Napolitano, M., Franco, R., Botti, G., and Castello, G. 2005. Expression of CXCR4 predicts poor prognosis in patients with malignant melanoma. *Clin. Cancer Res.* 11: 1835.

115. Donninger, H., Bonome, T., Radonovich, M., Pise-Masison, C. A., Brady, J., Shih, J. H., Barrett, J. C., and Birrer, M. J. 2004. Whole genome expression profiling of advance stage papillary serous ovarian cancer reveals activated pathways. *Oncogene* 23: 8065.

116. Hibbs, K., Skubitz, K. M., Pambuccian, S. E., Casey, R. C., Burleson, K. M., Oegema, T. R., Thiele, J. J., Grindle, S. M., Bliss, R. L., and Skubitz, A. P. 2004. Differential gene expression in ovarian carcinoma: identification of potential biomarkers. *Am. J. Pathol.* 165: 397.

117. Meinhold-Heerlein, I., Bauerschlag, D., Hilpert, F., Dimitrov, P., Sapinoso, L. M., Orlowska-Volk, M., et al. 2005. Molecular and prognostic distinction between serous ovarian carcinomas of varying grade and malignant potential. *Oncogene* 24: 1053.

118. Santin, A. D., Zhan, F., Bellone, S., Palmieri, M., Cane, S., Bignotti, E., et al. 2004. Gene expression profiles in primary ovarian serous papillary tumors and normal ovarian epithelium: identification of candidate molecular markers for ovarian cancer diagnosis and therapy. *Int. J. Cancer* 112: 14.

119. Le Page, C., Provencher, D., Maugard, C. M., Quellet, V., and Mes-Masson, A. M. 2004. Signature of a silent killer: expression profiling in epithelial ovarian cancer. *Exp. Rev. Molec. Diagnostics* 4: 157.

120. Sawasaki, T., Shigemasa, K., Gu, L., Beard, J. B., and O'Brien, T. J. 2004. The transmembrane protease serine (TMPRSS3/TADG-12) D variant: a potential candidate for diagnosis and therapeutic intervention in ovarian cancer. *Tumor Biol.* 25: 141.

121. Lee, B. C., Cha, K., Avraham, S., and Avraham, H. K. 2004. Microarray analysis of differentially expressed genes associated with human ovarian cancer. *Int. J. Oncol.* 24: 847.

122. Lancaster, J. M., Dressman, H. K., Whitaker, R. S., Havrilesky, L., Gray, J., Marks, J. R., Nevins, J. R., and Berchuck, A. 2004. Gene expression patterns that characterize advanced stage serous ovarian cancers. *J. Soc. Gyn. Invest.* 11: 51.

123. Liou, L. S., Shi, T., Duan, Z. H., Sadhukhan, P., Der, S. D., Novick, A. A., Hissong, J., Skacel, M., Almasan, A., and DiDonato, J. A. 2004. Microarray gene expression profiling and analysis in renal cell carcinoma. *BMC Urology* 4: 9.

124. Ami, Y., Shimazui, T., Akaza, H., Uematsu, N., Yano, Y., Tsujimoto, G., and Uchida, K. 2005. Gene expression profiles correlate with the morphology and metastasis characteristics of renal cell carcinoma cells. *Oncol. Reps.* 13: 75.

125. Dal Cin, P. 2003. Genetics in renal cell carcinoma. *Curr. Opin. Urology* 13: 463.

126. Burczynski, M. E., Twine, N. C., Dukart, G., Marshall, B., Hidalgo, M., Stadler, W. M., et al. 2005. Transcriptional profiles in peripheral blood mononuclear cells prognostic of clinical outcomes in patients with advanced renal cell carcinoma. *Clin. Cancer Res.* 11: 1181.

127. Li, W., Kessler, P., and Williams, B. R. 2005. Transcript profiling of Wilms tumors reveals connections to kidney morphogenesis and expression patterns associated with anaplasia. *Oncogene* 24: 457.

128. Yang, X. J., Sugimura, J., Tretiakova, M. S., Furge, K., Zagaja, G., Sokoloff, M., et al. 2004. Gene expression profiling of renal medullary carcinoma: potential clinical relevance. *Cancer* 100: 976.

129. Kim, J. M., Sohn, H. Y., Yoon, S. Y., Oh, J. H., Yang, J. O., Kim, J. H., et al. 2005. Identification of gastric cancer-related genes using a cDNA microarray containing novel expressed sequence tags expressed in gastric cancer cells. *Clin. Cancer Res.* 11: 473.

130. Chen, X., Leung, S. Y., Yuen, S. T., Chu, K. M., Ji, J., Li, R., et al. 2003. Variation in gene expression patterns in human gastric cancers. *Molec. Biol. Cell* 14: 3208.

131. Terashima, M., Maesawa, C., Oyama, K., Ohtani, S., Akiyama, Y., Ogasawara, S., et al. 2005. Gene expression profiles in human gastric cancer: expression of maspin correlates with lymph node metastasis. *Brit. J. Cancer* 92: 1130.

132. Leung, S. Y., Yuen, S. T., Chu, K. M., Mathy, J. A., Li, R., Chan, A. S., Law, S., Wong, J., Chen, X., and So, S. 2004. Expression profiling identifies chemokine (C-C motif) ligand 18 as an independent prognostic indicator in gastric cancer. *Gastroenterology* 127: 457.

133. Went, P. T., Dirnhofer, S., Bundi, M., Mirlacher, M., Schraml, P., Mangialaio, S., Dimitrijevic, S., Kononen, J., Lugli, A., Simon, R., and Sauter G. 2004. Prevalence of KIT expression in human tumors. *J. Clin. Oncol.* 22: 4515.

134. Dyrskjot, L. 2003. Classification of bladder cancer by microarray expression profiling: towards a general clinical use of microarrays in cancer diagnostics. *Exp. Rev. Molec. Diagnostics* 3: 635.

135. Dyrskjot, L., Thykjaer, T., Kruhoffer, M., Jensen, J. L., Marcussen, N., Hamilton-Dutoit, S., Wolf, H., and Orntoft, T. F. 2003. Identifying distinct classes of bladder carcinoma using microarrays. *Nat. Genetics* 33: 90.

136. Sanchez-Carbayo, M., and Cordon-Cardo, C. 2003. Applications of array technology: identification of molecular targets in bladder cancer. *Brit. J. Cancer* 89: 2172.

137. Sanchez-Carbayo, M., Socci, N. D., Lozano, J. J., Li, W., Charytonowicz, E., Belbin, T. J., Prystowsky, M. B., Ortiz, A.,R., Childs, G., and Cordon-Cardo, C. 2003. Gene discovery in bladder cancer progression using cDNA microarrays. *Am. J. Pathol.* 163: 505.

138. Nicholson, B. E., Frierson, H. F., Conaway, M. R., Seraj, J. M., Harding, M. A., Hampton, G. M., and Theodorescu, D. 2004. Profiling the evolution of human metastatic bladder cancer. *Cancer Res.* 64: 7813.

139. Lossos, I. S., and Levy, R. 2003. Diffuse large B-cell lymphoma: insights gained from gene expression profiling. *Int. J. Hematol.* 77: 321.

140. Wiestner, A., and Staudt, L. M. 2003. Towards molecular diagnosis and targeted therapy of lymphoid malignancies. *Semin. Hematol.* 40: 296.

141. Hirakawa, S., Hong, Y. K., Harvey, N., Schacht, V., Matsuda, K., Libermann, T., and Detmar, M. 2003. Identification of vascular lineage-specific genes by transcriptional profiling of isolated blood vascular and lymphatic endothelial cells. *Am. J. Pathol.* 162: 575.

142. Mischel, P. S., Nelson, S. F., and Cloughesy, T. F. 2003. Molecular analysis of glioblastoma: pathway profiling and its implications for patient therapy. *Cancer Biol. Therapy* 2: 242.

143. Kees, U. R., Ford, J., Watson, M., Murch, A., Ringner, M., Walker, R. L., and Meltzer, P. 2003. Gene expression profiles in a panel of childhood leukemia cell lines mirror critical features of the disease. *Molec. Cancer Therap.* 2: 671.

144. Shai, R., Shi, T., Kremen, T. J., Horvath, S., Liau, L. M., Cloughesy, T. F., Mischel, P. S., and Nelson, S. F. 2003. Gene expression profiling identifies molecular subtypes of gliomas. *Oncogene* 22: 4918.

145. Wright, A., Tan, B., Rosenwald, A., Hurt, E. H., Wiestner, A., and Staudt, L. M. 2003. A gene expression-based method to diagnose clinically distinct subgroups of diffuse large B-cell lymphoma. *Proc. Natl. Acad. Sci. USA* 100: 9991.

146. Skubitz, K. M., and Skubitz, A. P. 2003. Differential gene expression in leiomyosarcoma. *Cancer* 98: 1029.

147. Bruchova, H., Kalinova, M., and Brdicka, R. 2004. Array-based analysis of gene expression in childhood acute lymphoblastic leukemia. *Leukemia Res.* 28: 1.

148. Wang, J., Coombes, K. R., Highsmith, W. E., Keating, M. J., and Abruzzo, L. V. 2004. Differences in gene expression between B-cell chronic lymphocytic leukemia and normal B cells: a meta-analysis of three microarray studies. *Bioinformatics* 20: 3166.

149. Cole, K., Truong, V., Barone, D., and McGall, G. 2004. Direct labeling or RNA with multiple biotins allows sensitive expression profiling of acute leukemia class predictor genes. *Nucl. Acids Res.* 32: 86.

150. Wei, J. S., Greer, B. T., Westermann, F., Steinberg, S. M., Son, C. G., Chen, Q. R., et al. 2004. Prediction of clinical outcome using gene expression profiling and artificial neural networks for patients with neuroblastoma. *Cancer Res.* 64: 6883.

151. Algire, G. H., Chalkley, H. W., Legallais, F. Y., and Park, H. D. 1945. Vascular reactions of normal and malignant tumors in vivo. I. Vascular reactions of mice to wounds and to normal and neoplastic transplants. *J. Natl. Cancer Inst.* 6: 73.

152. Folkman, J., Merler, E., Abernathy, C., and Williams, G. 1971. Isolation of a tumor factor responsible for angiogenesis. *J. Exp. Med.* 133: 275.

153. Folkman, J. 1974. Tumor angiogenesis. *Adv. Cancer Res.* 19: 331.

154. Folkman, J., and Cotran, R. 1976. Relation of vascular proliferation to tumor growth. *Int. Rev. Exp. Pathol.* 16: 207.

155. Folkman, J. 1971. Tumor angiogenesis: therapeutic implications. *New Engl. J. Med.* 285: 1182.

156. Ferrara, N., and Davis-Smyth, T. 1997. Biology of vascular endothelial growth factor. *Endocr. Rev.* 18: 4.

157. Dvorak, H. F. 1986. Tumors: wounds that do not heal—similarities between tumor stroma generation and wound healing. *New Engl. J. Med.* 315: 1650.

158. Dvorak, H. F., Brown, L. F., Detmar, M., and Dvorak, A. M. 1995. Vascular permeability factor/vascular endothelial growth factor, microvascular hyperpermeability and angiogenesis. *Am. J. Pathol.* 146: 1029.

159. Dvorak, H. F., Harvey, V. S., Estrella, P., Brown, L. F., McDonagh, J., and Dvorak, A. M. 1987. Fibrin containing gels induce angiogenesis: implications for tumor stroma generation and wound healing. *Lab. Invest.* 57: 673.

160. Leung, D. W., Cachianes, G., Kuang, W. J., Goeddel, D. V., and Ferrara, N. 1989. Vascular endothelial growth factor is a secreted angiogenic mitogen. *Science* 246: 1306.

161. Jakeman, L. B., Armanini, M., Philips, H. S., and Ferrara, N. 1993. Developmental expression of binding sites and mRNA for vascular endothelial growth factor suggests a role for this protein in vasculogenesis and angiogenesis. *Endocrinology* 133: 848.

162. Breier, G., Albrecht, U., Sterrer, S., and Risau, W. 1992. Expression of vascular endothelial growth factor during embryonic angiogenesis and endothelial cell differentiation. *Development* 114: 521.

163. Shifren, J. L., Doldi, N., Ferrara, N., Mesiano, S., and Jaffe, R. B. 1994. In the human fetus, vascular endothelial growth factor is expressed in epithelial cells and myocytes, but not vascular endothelium: implications for mode of action. *J. Clin. Endocrinol. Metab.* 79: 316.

164. Ferrara, N., Houck, K., Jakeman, L., and Leung, D. W. 1992. Molecular and biological properties of the vascular endothelial growth factor family of proteins. *Endocr. Rev.* 13: 18.

165. Volm, M., Koomagi, R., Mattern, J., and Stammler, G. 1997. Angiogenic growth factors and their receptors in non-small cell lung carcinomas and their relationships to drug response in vitro. *Anticancer Res.* 17: 99.

166. Volm, M., Koomagi, R., and Mattern, J. 1997. Prognostic value of vascular endothelial growth factor and its receptor Flt-1 in squamous cell lung cancer. *Int. J. Cancer* 74: 64.

167. Brown, L. F., Berse, B., Jackman, R. W., Tognazzi, K., Guidi, A. J., Dvorak, H. F., et al. 1995. Expression of vascular permeability factor (vascular endothelial growth factor) and its receptors in breast cancer. *Hum. Pathol.* 26: 86.

168. Suzuki, K., Hayashi, N., Miyamoto, Y., Yamamoto, M., Ohkawa, K., Ito, Y., et al. 1996. Expression of vascular permeability factor/vascular endothelial growth factor in human hepatocellular carcinoma. *Cancer Res.* 56: 3004.

169. Guidi, A. J., Abu Jawdeh, G., Tognazzi, K., Dvorak, H. F., and Brown, L. F. 1996. Expression of vascular permeability factor (vascular endothelial growth factor) and its receptors in endometrial carcinoma. *Cancer* 78: 454.

170. Hashimoto, M., Ohsawa, M., Ohnishi, A., Naka, N., Hirota, S., Kitamura, Y., and Aozasa, K. 1995. Expression of vascular endothelial growth factor and its receptor mRNA in angiosarcoma. *Lab. Invest.* 73: 859.

171. Viglietto, G., Romano, A., Maglione, D., Rambaldi, M., Paoletti, I., Lago, C. T., et al. 1996. Neovascularization in human germ cell tumors correlates with a marked increase in the expression of the vascular endothelial growth factor but not the placenta-derived growth factor. *Oncogene* 13: 577.

172. Wizigmann Voos, S., Breier, G., Risau, W., and Plate, K. H. 1995. Up-regulation of vascular endothelial growth factor and its receptors in von Hippel-Lindau disease-associated and sporadic hemangioblastomas. *Cancer Res.* 55: 1358.

173. Ferrara, N., Winer, J., Burton, T., Rowland, A., Siegel, M., Phillips, H. S., et al. 1993. Expression of vascular endothelial growth factor does not promote transformation but confers a growth advantage in vivo to Chinese hamster ovary cells. *J. Clin. Invest.* 91: 160.

174. Qu, H., Nagy, J. A., Senger, D. R., Dvorak, H. F., and Dvorak, A. M. 1995. Ultrastructural localization of vascular permeability factor/vascular endothelial growth factor (VPF/VEGF) to the albuminal plasma membrane and vesiculovacuolar organelles of tumor microvascular endothelium. *J. Histochem. Cytochem.* 43: 381.

175. Ferrara, N., Hillan, K. J., Gerber, H.-P., and Novotny, W. 2004. Discovery and development of bevacizumab, an anti-VEGF antibody for treating cancer. *Nature Rev. Drug Discovery* 3: 391.

176. Taylor, K. L., Henderson, A. M., and Hughes, C. C. 2002. Notch activation during endothelial cell network formation in vitro targets the basic HLH transcription factor HESR-1 and downregulates VEGFR-2/KDR expression. *Microvascular Res.* 64: 372.

177. Ferrara, N., et al. 1996. Heterozygous embryonic lethality induced by targeted inactivation of the VEGF gene. *Nature* 380: 439.

178. Cameliet, P., et al. 1996. Abnormal blood vessel development and lethality in embryos lacking a single VEGF allele. *Nature* 380: 435.

179. Gerber, H. P., et al. 1999. VEGF is required for growth and survival in neonatal mice. *Development* 126: 1149.

180. Ravindranath, N., Little-Ihrig, L., Phillips, H. S., Ferrara, N., and Zeleznik, A. J. 1992. Vascular endothelial growth factor messenger ribonucleic acid expression in the primate ovary. *Endocrinology* 131: 254.

181. Ferrara, N., et al. 1998. Vascular endothelial growth factor is essential for corpus leteum angiogenesis. *Nature Med.* 4: 336.

182. Hazzard, T. M., Xu, F., and Stouffer, R. L. 2002. Injection of soluble vascular endothelial growth factor receptor 1 into the preovulatory follicle disrupts ovulation and subsequent luteal function in rhesus monkeys. *Biol. Reprod.* 67: 1305.

183. Zachary, I., and Gliki, G. 2001. Signaling transduction mechanisms mediating biological actions of the vascular endothelial growth factor family. *Cardiovascular Res.* 49: 568.

184. Tjwa, M., Luttun, A., Autiero, M., and Carmeliet, P. 2003. VEGF and PlGF: two pleio-tropic growth factors with distinct roles in development and homeostasis. *Cell Tissue Res.* 314: 5.

185. Prior, B. M., Lloyd, P. G., Yang, H. T., and Terjung, R. L. 2003. Exercise-induced vas-cular remodeling. *Exercise Sport Rev.* 31: 26.

186. Dor, Y., Djonov, V., and Keshet, E. 2003. Induction of vascular networks in adult organs: implications to proangiogenic therapy. *Ann. N.Y. Acad. Sci.* 995: 208.

187. Bates, D. O., and Harper, S. J. 2002. Regulation of vascular permeability by vascular endothelial growth factors. *Vascular Pharmacol.* 39: 225.

6 RNA Viruses and RNA-Based Drugs
A Perfect Match for RNA Delivery and the Identification of Candidate Therapeutic Target Inflammatory Molecules

Brett A. Lidbury, Cristina M. Musso, Jasjit Johal, Nestor E. Rulli, and Suresh Mahalingam
Virus and Inflammation Research Group,
Centre for Biomolecular and Chemical
Sciences, University of Canberra

Mark T. Heise
The Carolina Vaccine Institute, Department of Genetics,
and Department of Microbiology and Immunology,
University of North Carolina at Chapel Hill

CONTENTS

6.1 INTRODUCTION

Viruses have an unmatched genius at infiltrating cells and manipulating the expression of host genes that influence viral replication and survival. This complexity of virus–host interaction presents significant challenges to contemporary medical science, with viral diseases continuing to cause mortality and morbidity worldwide. Viruses of current major concern to human health show the majority to be those comprising an RNA genome (e.g., HIV, influenza, dengue) [1]. Such RNA virus capacity in terms of the manipulation of host-cell function, in terms of potential RNA-based gene therapies and vaccines, does have positive dimensions:

1. The ability of such viruses to infiltrate cell membranes and deliver genetic material to the cell interior, and thereafter exploit the cell's molecular machinery to replicate viral genes
2. The complexities of the virus–host relationship that emphasize the crucial host molecules/pathways responsible for antiviral defense, as well as inflammation control.

This chapter focuses on two aspects learned from RNA viruses:

1. A review of RNA viral vectors (e.g., alphaviral "replicons") and their potential to deliver therapeutic RNA molecules to individuals suffering from a disease
2. The identification of host inflammatory pathways and immune responses that, because of their potent antiviral activities, are targeted by viruses for disruption, evasion, or sabotage.

The combination of knowledge in gene delivery and in RNA virus-manipulated host defense responses provides an exciting future arena for RNA-based drug targets, as well as for gene-therapy vector designs derived from engineering RNA virus genomes.

6.2 RNA VIRUS VECTORS AND REPLICON TECHNOLOGIES

Since the demonstration by Racaniello and Baltimore in 1981 that RNA from a molecularly cloned poliovirus was infectious [2], there has been significant interest in utilizing RNA viruses as vectors for vaccination or gene therapy. For many years, much of this work was limited to positive-sense RNA viruses, such as the picornaviruses, togaviruses, and flaviviruses, due to the development of infectious cDNA clones for many of these viruses [2–4]. However, in recent years, infectious cDNA clones have been generated for negative-sense RNA viruses, as well as the large coronaviruses, raising the possibility of generating expression vectors based on many different virus types.

RNA virus vectors, which for the purpose of this discussion are viruses or their derivatives that express heterologous genetic material that has been engineered into the viral genome, can be classified either as replication-competent viral vectors or replication-defective "replicon vectors." Replication-competent vectors are fully functional viruses that are capable of infecting a cell and producing progeny virions that will infect subsequent cells. However, a heterologous genetic sequence, which can be as small as the coding sequence of a T-cell epitope or several thousand nucleotides in length, is also expressed from the viral genome. Nonreplication-competent vectors are generally lacking one or more viral genes that are essential for the production of progeny virions, with the heterologous gene expressed in place of the missing viral sequences. Commonly, the missing genes are viral structural genes, which encode the proteins that physically make up the virion particle, while the genes encoding the viral replicase proteins, which mediate viral RNA synthesis, remain intact. Therefore, if the RNA for the defective vector is introduced into a permissive cell, viral RNA synthesis will commence, but the lack of structural genes will prevent the production of progeny virions. Though these vectors can be used *in vitro* by simply introducing the defective viral genome RNA into a permissive cell, it is often possible to supply the viral structural genes in *trans*, which will allow the defective genome to be packaged, with the resulting defective viral particles capable of only a single round of replication.

6.2.1 REPLICATION-COMPETENT VIRAL VECTORS

Replication-competent viral expression vectors have been constructed for a number of virus families, including togaviruses and coronaviruses [5, 6]. The major use, or intended use, of these vectors is in vaccination, where the engineered virus expressed heterologous nucleic acid sequences ranging in size from T- and B-cell epitopes to entire open reading frames. Three basic strategies have been employed to express heterologous genetic material from replication-competent vectors. The first involves the insertion of short genetic sequences into viral genes that will tolerate the additional sequence. This approach has been successfully applied to alphaviruses, where antigenic sequences from heterologous pathogens have been engineered into the glycoproteins of Sindbis virus [7]. A second approach is to replace a viral gene that is nonessential for viral replication with a heterologous sequence. For example, for coronaviruses, the transmissible gastroenteritis virus (TGEV) genome has been engineered to express heterologous genes in place of the nonessential 3a and 3b open reading frames [8], while in the mouse hepatitis virus genome, the nonessential gene 4 can be replaced with a heterologous gene [9]. Finally, a genetic sequence can be expressed from the viral genome by engineering in a new viral promoter to drive expression of the heterologous sequence. This approach has been used extensively with alphaviruses, where additional copies of the 26S subgenomic promoter can be introduced at the beginning of the viral 3′ UTR (untranslated region). This additional subgenomic promoter will then drive high-level expression of the heterologous sequence in infected cells [10].

There are several advantages and disadvantages to using replication-competent expression vectors. One major advantage with replication-competent vectors is that they can be propagated in culture by simply allowing the vector to grow, which greatly decreases the cost of producing/packaging the vector. The ability of the vector to spread and replicate within the inoculated host may also result in the generation of more robust immune responses against the vector and the protein derived from the heterologous sequence. However, though many virus-based expression vectors are derived from existing live attenuated vaccines, such as the 17D strain of yellow fever virus [4], there are obviously some safety concerns whenever a replication-competent vector is used, especially in highly susceptible populations, such as immunocompromised individuals. Many vectors also have limitations with respect to how large a sequence can be inserted, with some vectors limited to the coding sequence for one or a few B- or T-cell epitopes. Finally, vector stability is also an issue, where many recombinant viruses lose transgene expression after a few rounds of replication [11].

6.2.2 PROPAGATION-DEFECTIVE VIRAL VECTORS

In addition to replication-competent viral vectors, propagation-defective viral vectors, or "replicons," have been developed for a number of RNA viruses. Replicon vectors are generally made by removing one or more of the viral structural genes from the viral genome, which results in a virus that can initiate viral RNA synthesis and gene expression, but is unable to produce infectious progeny virions. The missing gene or genes can then be provided in *trans* to package the viral RNA, with the resulting "single-hit" particle (replicon particle) able to infect a new target cell and initiate viral RNA synthesis and gene expression, but not produce progeny virion due to the lack of virally encoded structural genes. This approach has been used effectively with a number of positive-sense RNA viruses, including picornaviruses, togaviruses, flaviviruses, and coronaviruses, as well as negative-sense RNA viruses such as vesicular stomatitis virus (VSV) [12–15].

Alphavirus vectors provide an excellent example of the power of replicon vectors for gene delivery. Though the genome is a single-stranded RNA of positive-sense polarity, the genome can readily be broken down into two functional units. The 5′ two-thirds of the viral genome encodes the viral nonstructural genes (nSP), which are essential for viral RNA synthesis, while the 3′ one-third of the viral genome encodes the viral structural genes, which are expressed from a highly active subgenomic promoter. Therefore, the structural genes, which are made up of a capsid gene, two major glycoproteins (E2 and E1), as well as two additional proteins (E3 and 6K) that serve as signal sequences, can be removed from the genome without affecting viral RNA synthesis. A heterologous gene can then be expressed from the subgenomic promoter in place of the structural genes. The resulting replicon RNA will initiate viral RNA synthesis if introduced into a permissive cell, and drive high-level expression of the heterologous gene, though no progeny virions will be produced [16]. The replicon RNA can be packaged by providing helper RNA, which contains the alphavirus 5′ and 3′ ends, as well as the subgenomic promoter driving expression of the viral structural genes, but which have deletion of the viral nonstructural genes [17]. If this helper RNA is introduced into the same cell as the replicon RNA, the replicon-derived nonstructural proteins will amplify the helper RNA in *trans*, and drive expression of the viral structural genes from the helper. The structural proteins will then package the replicon RNA to produce replicon particles, while the helper RNA, which lacks an RNA packaging signal [16], will not be packaged, or will be packaged inefficiently. In order to decrease the likelihood of a recombination event between the helper and the replicon RNA that would generate a replication-competent virus, a split helper system, where the viral capsid is expressed from one helper RNA and the viral glycoproteins are expressed from a second helper, can also be used [17]. Variations on this theme, including packaging cell lines that stably express the alphavirus structural genes [17], have also been used to produce alphavirus replicon particles. Vectors based on several different alphaviruses have been used for vaccine delivery or in short-term gene therapy applications.

Single-hit vectors have also been constructed for a number of negative-sense RNA viruses, including VSV. Defective VSV vectors can be generated by deletion of the viral G protein, which is essential for the production of infectious viruses [15]. Heterologous genes can be inserted in place of G within the VSV genome, and the viral G protein then provided in *trans* using a packaging cell line transduced with a mammalian expression vector driving VSV G expression. One advantage of these vectors is the fact that the defective VSV particles can be pseudotyped with glycoproteins from a wide range of viruses, which will allow the vectors to be directed to specific cell types, depending on which viral glycoprotein is used for packaging [15].

6.3 RNA VIRUS EVASION OF THE INTERFERON SYSTEM

It is widely published that DNA viruses utilize a number of strategies to evade the antiviral immune responses of the host [18]. DNA viruses are thought to have "stolen" genes from the host that were modified to benefit virus survival [18]. However, fewer examples exist for smaller-genome RNA viruses, and information on immune evasion strategies employed by this class of viruses is emerging [1]. Several important immune evasion strategies employed by RNA viruses are discussed below for the human immunodeficiency virus (HIV), influenza virus, and Ross River virus (RRV), with a special focus on the interferon (IFN) system. The type I interferons are crucial in exerting antiviral effects against RNA viruses. Understanding the various IFN evasion mechanisms employed by these viruses will assist in the development of appropriate drugs or other therapeutic strategies in combating these and other virus pathogens.

The activation of host proteins such as IFNs, following viral infection, appears to correlate with early defense mechanisms, since IFNs act to inhibit both DNA and RNA viruses. Because of this, viruses often target IFNs in their attempt to modify or neutralize their antiviral potency [19]. The antiviral effects of IFNs are mediated through various intracellular pathways and are initiated by their binding to cellular receptors, which are present on most cells [20] (Figure 6.1). For instance, transcription factors such as STAT (signal transducer and activator of transcription) complexes become phosphorylated following binding of type I IFNs to its receptor. This is followed by the translocation of phosphorylated STAT complexes (consisting of STAT-1 and STAT-2 heterodimers) to the nucleus, where they bind to transcription elements of several antiviral genes (IFN-β, RNA-dependent protein kinase [PKR]), 2′ 5′ oligoadenylate synthetase [OAS], nitric oxide [NO], as well as secondary transcription factors [interferon-regulatory factor 1 [IRF-1], IRF-3, IRF-7]]. Genes such as IRF-1, IRF-3, and IRF-7, which become activated, also play important roles in the transcription of various other downstream antiviral genes. The production of these antiviral proteins

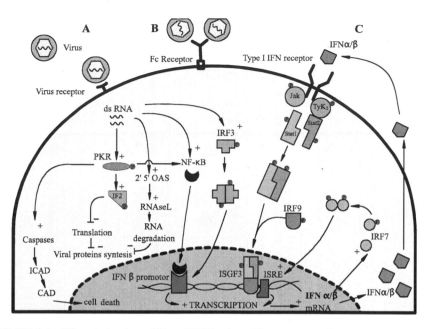

FIGURE 6.1   Virus induction of Type I IFNs. A and B represent primary events. C represents host response. Arrows (→) indicate stimulation or induction. Dashed lines (–) indicate suppression or inhibition. Virus infection and double-stranded RNA (dsRNA) will trigger the early activation of different cell proteins (PKR, NF-κB, 2′5′ OAS, IRF3) that mediate the primary expression of IFN alpha/beta. Type I IFN interaction with the specific cell receptor stimulates Jak1 and Tyk2 proteins. The activation of STAT-1 and STAT-2 is followed by the formation of an ISGF3 (interferon-stimulated gene factor-3) complex. ISGF3 will bind to the interferon receptor ISRE, promoting IFN synthesis and IRF-7 induction.

can directly disrupt the regulation of viral and cellular macromolecular synthesis and degradation. In recent years, many human RNA viruses have been shown to interrupt/disrupt the IFN system by targeting various proteins in the signaling cascade, such as STAT complexes, IRF proteins, 2′5′ OAS, and PKR [1, 19, 20].

6.3.1   HUMAN IMMUNODEFICIENCY VIRUS

Human Immunodeficiency virus (HIV) is of major concern worldwide, with devastating health impacts seen particularly in the developing world. HIV infection leads to the condition known as acquired immunodeficiency syndrome (AIDS), and is transmitted primarily through infected blood or semen. HIV is extremely successful in modulating various compartments of the host immune responses to its benefit. In the past 10 years, studies have shown that several HIV proteins, such as Nef, Env, and Vpu, can cripple the host immune responses. These proteins are shown to abolish CD4 surface protein from infected cells [21], resulting in the inability of T cells to engage with antigen-presenting cells and mount an immune response. Additionally, the Nef protein can trigger the down-regulation of MHC-I molecules, suppress apoptosis of infected cells by inhibiting the apoptosis-associated kinase 1 (ASK 1), as well as stimulate the expression of Fas ligands on T cells, inducing their death [22].

In addition, immune regulatory processes may also be manipulated through the HIV-1 Tat protein, a potent monocyte chemoattractant expressed early in the life cycle of this virus [23]. Tat has been shown to facilitate infection by the recruitment of mononuclear cells toward HIV-producing cells, possible through its expression of conserved amino acid sequences that correspond to critical chemokine sequences. Furthermore, it is possible that Tat plays a role in the down-regulation of both HLA class I and II molecules, although there is much controversy present within the literature [24–26].

The binding of the virus to a specific cell receptor, viral cellular membrane fusion, and dsRNA activates NF-κB, resulting in a strong transcriptional stimulation of several early viral genes. Tat has also been shown to inhibit HIV-1 Tar (transactivation response) RNA binding and activation of 2′-5′ OAS. The HIV-1 Tat and Tar RNA binding proteins were shown to inhibit dsRNA-mediated activation of PKR [27]. Tar-RNA forms an inactive heterodimer with PKR. Tat, on the other hand, inhibits autophosphorylation of PKR and competes with eIF2α [28, 29] (Figure 6.2).

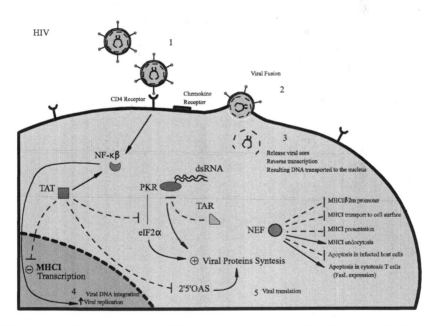

FIGURE 6.2 Early events after HIV infection of a host cell. After internalization and uncoating, reverse transcription results in the production of double-stranded DNA. The viral DNA is integrated into the host genome. Transcription results in various viral mRNAs encoding the regulatory and structural proteins to be translated in the cytosol (steps 1–5). The binding of the virus to a specific cell receptor, viral cellular membrane fusion, and dsRNA activates NF-κB, resulting in a strong transcriptional stimulation of several early viral genes. TAR and TAT are known to down-regulate PKR function. TAR-RNA forms an inactive heterodimer with PKR. TAT inhibits autophosphorylation of PKR and competes with eIF2α. NEF induces down-regulation of MHC I molecules and inhibits apoptosis of the infected cells by inhibiting the apoptosis-associated kinase 1 (ASK 1), but stimulates the expression of Fas ligands on the surface of T cells, inducing their death.

Other research has demonstrated the ability of HIV to manipulate the chemokine system. It was shown that chemokine receptors were utilized as co-receptors to facilitate entry into CD4+ cells [30]. Activity of cells expressing CD4 as well as CCR5 and CXCR4 may be influenced by the HIV structural protein, gp120, and benefited by its interaction with cellular receptors for viral entry. Such binding activity by gp120 can elicit not only apoptosis in CD4+ T cells and endothelial cells, but other outcomes affecting function, including neurodegenerative effects and dysregulated lymphocyte homing [31].

6.3.2 INFLUENZA VIRUS

The influenza virus continues to pose significant global health problems, causing morbidity and mortality worldwide. In the United States, mortality rates over the last 30 years have ranged between 30,000 to 70,000 cases per year, with the elderly and those who are immunocompromised most at risk of death after influenza virus infection [32].

Insights into influenza pathogenesis have been helped over the past decade by studies into the modulation of the host type I IFN response via influenza proteins postinfection. The nonstructural protein of influenza A, NS1, has been shown to have a role in inhibiting the host activation of type I IFNs [33, 34], associated with the disruption of PKR activation by NS1 binding to dsRNA [35]. NF-κB and other transcription factors play a role in the transactivation of the IFN-β promoter, and in addition to the capacity for influenza to block the PKR response, NS1 can inhibit the activation of both NF-κB and IRF-3 [36, 37], thus compromising transcriptional activation pathways associated with IFN expression (see Figure 6.3). Therefore, it is evident that NS1 has an immune evasion role in addition to its polymerase activity, suggesting that viral proteins are capable of executing diverse functions within the viral life cycle. For these and other studies, an influenza virus was engineered with a deleted NS1 gene (delNS1), and early results showed that the delNS1 influenza virus could only replicate in IFN-deficient cells [33]. In IFN-intact cells, delNS1 influenza virus infection stimulated increased type I IFN mRNA transcription [33, 34], as well as high levels of NF-κB activation, in comparison to the NS1-competent parent virus [37], which suggests that the growth inhibition of delNS1 influenza was a result of the robust IFN response postinfection.

Subsequent work on influenza virus interaction with the type I IFN response has also revealed some of the subtle mechanisms employed by the virus to disrupt the IFN antiviral effect. As discussed earlier, influenza infection disrupts PKR activity in the infected host cell via the viral NS1 protein, allowing enhanced influenza virus replication [35]. The early IFN response to influenza infection has found a critical role for the host PKR inhibitory protein, P58[IPK], which, through binding to the PKR kinase domain, leads to the suppression of PKR-mediated phosphorylation of eIF2α [38, 39], a key intracellular protein required for both host and viral protein translation. P58[IPK] is bound to I-P58[IPK] as an inactive complex under conditions free of cell stress, such as infection. However, following infection with influenza virus, this inactive P58[IPK]/I-P58[IPK] complex becomes disrupted, and P58[IPK] is released to interact with PKR and inhibit its kinase activity. Therefore, with decreased phosphorylation

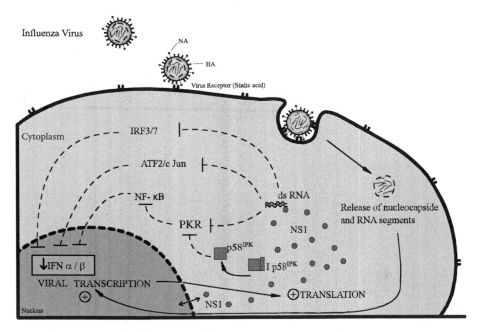

Influenza Virus

NA

HA

Virus Receptor (Sialic acid)

Cytoplasm

IRF3/7

ATF2/c Jun

NF- κB

PKR

ds RNA

Release of nucleocapside and RNA segments

NS1

p58^{IPK}

I p58^{IPK}

↓IFN α / β

VIRAL TRANSCRIPTION

⊕TRANSLATION

NS1

Nucleus

FIGURE 6.3 Interference of the interferon pathway by the influenza virus. Influenza virus infection results in expression of NS1 protein. NS1-dsRNA binding has the ability to prevent IRF3/7, ATF2/cJUN, NF-κB, and PKR activation, resulting in the inhibition of IFN α/β. The virus mobilizes a cellular protein (p58^{IPK}), which is released from its own inhibitor I-p58^{IPK} to repress PKR activity. Full lines indicate induction; dotted lines indicate inhibition.

of eIF2α by PKR, the translation of influenza virus proteins is increased, leading to the enhanced production of progeny influenza virus by the infected cell [38].

6.3.3 ROSS RIVER VIRUS (RRV)

Ross River virus (RRV) is an endemic Australian alphavirus and the agent responsible for the greatest incidence of arboviral disease in Australia. Disease resulting from infection is not fatal, but involves a syndrome of symptoms that includes arthritis/arthralgia, myalgia, lethargy, and/or rash, which can persist for several weeks to months [40]. RRV is closely related to the reemergent Chikungunya virus (CHIK; *Alphaviridae*, Semliki Forest virus clade) [41], which has recently caused dramatic disease outbreaks in the French territories of the West Indian Ocean, as well as in southern India [42]. RRV and CHIK infections have similar disease manifestations in humans (arthritis, lethargy, etc.), and along with other relevant members of the Alphavirus family are broadly referred to as "arthrogenic/arthritogenic alphaviruses" [40]. Macrophage and monocyte infiltrates have been associated with human disease [43], and monocyte/macrophage (F4/80+) cells have been identified as the cellular agent of severe muscle damage in RRV-infected mice [44]. Further studies in the mouse model have also detected inflammatory infiltrates in the joint and bone tissue of RRV-infected mice. Again, the infiltrate was predominantly F4/80+

cells, with significant populations of CD4+ T cells and NK cells detected via flow cytometric analysis [45].

RRV replicates in human and murine macrophages after cell uptake directly via a "natural" cellular receptor, or through Fc-γ receptor involving the "antibody-dependent enhancement" (ADE) mechanism of infection [46]. Since its discovery and implication as an important infection strategy, ADE has become a subject of significant scrutiny by many interested in human disease viruses. Essentially, the mechanism of ADE involves antibody-facilitated virus entry into macrophages or monocytes via Fc-γ receptors, resulting in a subsequent enhanced growth of RRV or other ADE viruses [46–48]. The first observation of RRV-ADE was reported in 1996 by Linn and colleagues [46], with the RRV-ADE and macrophage infection model leading to novel insights into the viral control of early immune and inflammatory host-gene expression postinfection.

An intriguing observation associated with RRV-ADE was the early suppression or ablation of macrophage antiviral responses, mediated by the disruption of TNF and NOS2 expression (mRNA and protein) [49]. In experiments involving pre-infection lipopolysaccharide (LPS)-stimulated antiviral activity for macrophage cultures (through enhanced pro-inflammatory cytokine and IFN expression), it was found that RRV was able to undergo unrestricted replication where infection was facilitated by polyclonal anti-RRV antibodies at a subneutralizing titer [49]. Conditions of considerable antiviral activity induced by LPS in macrophages did not allow RRV to grow when infection occurs through non-ADE (direct) infection [49]. The decrease in expression of TNF and NOS2 after RRV-ADE infection was associated with the disruption of the transcription factors interferon regulatory factor-1 (IRF-1) and nuclear factor-κB (NF-κB) [49, 50]. Significantly, the transcription and translation of non-antiviral *de novo* cellular proteins were unaffected by RRV-ADE [49]. Additional investigations showed that RRV-ADE infection stimulated increased interleukin-10 (IL-10) expression in macrophage cultures [50], explaining the global suppression of inflammatory/antiviral proteins after RRV-ADE infection (see Figure 6.4). This study also showed the suppression of IFN-associated STAT transcriptional protein complexes (e.g., ISGF-3), while the IL-10 associated transcription factor, Sp-1, was not affected by RRV-ADE infection [50].

The increased expression of IL-10 after RRV-ADE infection, at the same time as the suppression/ablation of TNF, NOS2, and type I IFN, suggests that viral infection mediated by ADE enables a sophisticated manipulation of host defense responses involving antiviral cytokines and immune protein expression. It is not simply the case that infection results in either a general suppression or general enhancement of host-gene expression postinfection. This ability of viruses to manipulate the cellular response through their interaction with transcriptional pathways that regulate key defense gene expression will provide important clues to the future design of antiviral and antiinflammatory genetic drugs, for example, the identification of gene candidates for targeting, either to augment or ablate downstream expression, during a given virus infection. The desired outcome for such a strategy is, on one hand, to enhance host immunity for optimum virus clearance, while also reducing host pathology and associated patient side effects. Also, with the above example of TNF/IL-10 expression for RRV-ADE, future antiviral genetic drug design will benefit

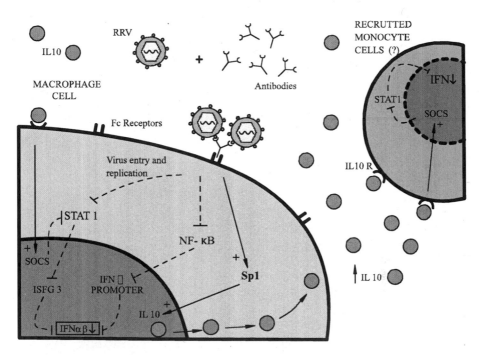

FIGURE 6.4 Suppression of antiviral response by RRV-ADE. RRV-ADE infection and replication results in the suppression of STAT-1 and NF-κB, but not Sp-1. As a consequence, the production of IFN α/β is inhibited while IL-10 is increased. The suppression of antiviral response through IL-10 mediated activity in monocyte cells is most likely mediated via up-regulation of SOCS3.

from transcriptional profiling of host genes during infection, allowing both the profiling of individual patients to determine disease severity and outcome, while also giving clues to the essential defense gene/protein interactions during infection for therapeutic targeting.

6.4 CONCLUDING REMARKS

With the exciting development of RNA-based gene technologies (for example, RNAi and anti-sense technologies) promising a revolution in drug and vaccine development, it is essential that leaders in such fields take pause to consider fundamental questions associated with pathogen-host interaction and disease mechanism with new drug development. The key questions involve the challenges of effective drug delivery and the development of drugs that, while curing a health disorder or providing disease protection, do not induce deleterious side effects in the patient. In this realm of new gene-based drug technologies, RNA viruses offer enormous hope that the above challenges can be met and overcome. As outlined in the introduction in section 6.1, viruses, through their fundamental nature and life cycle, have a genius in penetrating cells and delivering genetic material to the cell interior for expression. Furthermore, if constructed correctly through recombinant DNA technology, the virus will effectively deliver, replicate, and express therapeutic genes in target cells, leading to the subsequent modification of a biological response to ameliorate disease. In addition to these properties, and in the context of broader disease processes like inflammation, RNA viruses have proven to be outstanding tools in the elucidation of key host intracellular pathways that control aberrant immune and/or inflammatory response. In this regard, we have focused on the lessons learned from HIV and influenza virus, as well as Ross River virus antibody-dependent enhancement (RRV-ADE), which has exposed how the involvement of host-immunity can assist in the viral manipulation of the early response to infection.

In terms of the first challenge suggested above, viruses as natural gene delivery vectors provide a way to specifically target cells and tissues associated with a disease process. In terms of unwanted side effects, our knowledge of virus–host interaction will assist in the identification of key molecules and intracellular pathways that contribute to nonspecific inflammation and overzealous immune activity that can result from infection or other biological stimuli, and thereafter enhance or exacerbate disease. The delivery of exogenous genes via therapeutic virus infection will be used to block the overexpression of "disease" genes while also, where necessary, reconstituting critical host-gene activation pathways after specific suppression by viral proteins or nucleic acids.

The application of knowledge from the discipline of virology, both engineering RNA virus genomes and the molecular biology of virus–host interaction, is tailormade to accelerate developments in modern drug design using gene profiles from both the infected eukaryotic cell and the infecting virus itself.

ACKNOWLEDGMENTS

Suresh Mahalingam is an Australian NHMRC R.D. Wright Fellow.

REFERENCES

1. Mahalingam, S., Meanger, J., Foster, P. S., and Lidbury, B. A. 2002. The viral manipulation of the host cellular and immune environments to enhance propagation and survival: a focus on RNA viruses. *J. Leukocyte Biol.* 72: 429.
2. Racaniello, V. R., and Baltimore, D. 1981. Cloned poliovirus complementary DNA is infectious in mammalian cells. *Science* 214: 916.
3. Xiong, C., Levis, R., Shen, P., Schlesinger, S., Rice, C. M., and Huang, H. V. 1989. Sindbis virus: an efficient, broad host range vector for gene expression in animal cells. *Science* 243: 1188.
4. Rice, C. M., Grakoui, A., Galler, R., and Chambers, T. J. 1989. Transcription of infectious yellow fever RNA from full-length cDNA templates produced by in vitro ligation. *New Biol.* 1: 285.
5. Smerdou, C., and Liljeström, P. 2000. Alphavirus vectors: from protein production to gene therapy. *Gene Therapy and Regulation* 1: 33.
6. Enjuanes, L., Sola, I., Alonso, S., Escors, D., and Zuniga, S. 2005. Coronavirus reverse genetics and development of vectors for gene expression. *Curr. Top. Microbiol. Immunol.* 287: 161.
7. London, S. D., Schmaljohn, A. L., Dalrymple, J. M., and Rice, C. M. 1992. Infectious enveloped RNA virus antigenic chimeras. *Proc. Natl. Acad. Sci. USA* 89: 207.

8. Sola, I., Alonso, S., Zuniga, S., Balasch, M., Plana-Duran, J., and Enjuanes, L. 2003. Engineering the transmissible gastroenteritis virus genome as an expression vector inducing lactogenic immunity. *J. Virol.* 77: 4357.

9. Fischer, F., Stegen, C. F., Koetzner, C. A., and Masters, P. S. 1998. Construction of a mouse hepatitis virus recombinant expressing a foreign gene. *Adv. Exp. Med. Biol.* 440: 291.

10. Frolov, I., Hoffman, T. A., Prágai, B. M., Dryga, S. A., Huang, H. V., Schlesinger, S., and Rice, C. M. 1996. Alphavirus-based expression vectors: strategies and applications. *Proc. Natl. Acad. Sci. USA* 93: 11371.

11. Caley, I. J., Betts, M. R., Davis, N. L., Swanstrom, R., Frelinger, J. A., and Johnston, R. E. 1999. Venezuelan equine encephalitis virus vectors expressing HIV-1 proteins: vector design strategies for improved vaccine efficacy. *Vaccine* 17: 3124.

12. Kaplan, G., and Racaniello, V. R. 1988. Construction and characterization of poliovirus subgenomic replicons. *J. Virol.* 62: 1687.

13. Bredenbeek, P. J., Frolov, I., Rice, C. M., and Schlesinger, S. 1993. Sindbis virus expression vectors: packaging of RNA replicons by using defective helper RNAs. *J. Virol.* 67: 6439.

14. Khromykh, A. A., and Westaway, E. G. 1997. Subgenomic replicons of the flavivirus Kunjin: construction and applications. *J. Virol.* 71: 1497.

15. Roberts, A., Buonocore, L., Price, R., Forman, J., and Rose, J. K. 1999. Attenuated vesicular stomatitis viruses as vaccine vectors. *J. Virol.* 73: 3723.

16. Pushko, P., Parker, M., Ludwig, G. V., Davis, N. L., Johnston, R. E., and Smith, J. F. 1997. Replicon-helper systems from attenuated Venezuelan equine encephalitis virus: expression of heterologous genes in vitro and immunization against heterologous pathogens in vivo. *Virology* 239: 389.

17. Polo, J. M., Belli, B. A., Driver, D. A., Frolov, I., Sherrill, S., et al. 1999. Stable alphavirus packaging cell lines for Sindbis virus and Semliki Forest virus-derived vectors. *Proc. Natl. Acad. Sci. USA.*, 96: 4598.

18. Alcami, A., and Koszinowski, U. H. 2000. Viral mechanisms of immune evasion. *Immunol. Today* 21: 447.

19. Vilcek, J., and Sen, I. C. 1996. Interferons and other cytokines. In *Field's Virology*, eds. B. N. Fields, D. N. Knipe, P. M. Howley, R. M. Chanock, J. L. Melnick, T. P. Monath, B. Roizman, and S. E. Straus. Philadelphia: Lippincott-Raven, p. 375.

20. Samuel, C.E. 2001. Antiviral actions of interferons. *Clin. Microbiol. Rev.* 14: 778.

21. Piguet, V., Schwartz, O., Le Gall, S., and Trono, D. 1999. The downregulation of CD4 and MHC-I by primate lentiviruses: a paradigm for the modulation of cell surface receptors. *Immunol. Rev.* 168: 51.

22. Xu, X. N., Laffert, B., Screaton, G. R., Kraft, M., Wolf, D., Kolanus, W., et al. 1999. Induction of Fas ligand expression by HIV involves the interaction of Nef with T cell receptor zeta chain. *J. Exp. Med.* 189: 1489.

23. Albini, A., Ferrini, S., Benelli, R., Sforzini, S., Giunciuglio, D., Aluigi, M. G., Proudfoot, A. E., et al. 1998. HIV-1 Tat protein mimicry of chemokines. *Proc. Natl. Acad. Sci. USA.* 95: 13153.

24. Howcroft, T. K., Strebel, K., Martin, M. A., and Singer, D. S. 1993. Repression of MHC class I gene promoter activity by two-exon Tat of HIV. *Science* 260: 1320.

25. Tosi, G., De Lerma Barbaro, A., D'Agostino, A., Valle, M. T., Megiovanni, A. M., Manca, F., et al. 2000. HIV-1 Tat mutants in the cysteine-rich region downregulate HLA class II expression in T lymphocytic and macrophage cell lines. *Eur. J. Immunol.* 30: 19.

26. Matsui, M., Warburton, R. J., Cogswell, P. C., Baldwin, A. S., Jr., and Frelinger, J. A. 1996. Effects of HIV-1 Tat on expression of HLA class I molecules. *J. Acquir. Immune Defic. Syndr. Hum. Retrovirol.* 11: 233.

27. Park, H., Davies, M. V., Langland, J. O., Chang, H. W., Nam, Y. S., Tartaglia, J., Paoletti, E., et al. 1994. TAR RNA-binding protein is an inhibitor of the interferon-induced protein kinase PKR. *Proc. Natl. Acad. Sci. USA* 91: 4713.

28. Hiscott, J., Kwon, H., and Genin, P. 2001. Hostile takeovers: viral appropriation of the NF-κB pathway. *J. Clin. Invest.* 107: 143.

29. Demarch, F., Gutierrez, M. I., and Giacca, M. 1999. Human Immunodeficiency Virus Type 1 Tat protein activates transcription factor NF-κB through the cellular interferon-inducible, double-stranded RNA-dependent protein kinase, PKR. *J. Virol.* 73: 7080.

30. Berger, E. A., Murphy, P. M., and Farber, J. M. 1999. Chemokine receptors as HIV-1 coreceptors: roles in viral entry, tropism, and disease. *Annu. Rev. Immunol.* 17: 657.

31. Huang, M. B., Hunter, M., and Bond, V. C. 1999. Effect of extracellular human immunodeficiency virus type 1 glycoprotein 120 on primary human vascular endothelial cell cultures. *AIDS Res. Hum. Retroviruses* 15: 1265.

32. Thompson, W. W., Comanor, L., and Shay, D. K. 2006. Epidemiology of seasonal influenza: use of surveillance data and statistical models to estimate the burden of disease. *J. Infect. Dis.* 194 (Suppl. 2): S82.

33. Garcia-Sastre, A., Egorov, A., Matassov, D., Brandt, S., Levy, D. E., Durbin, J. E., et al. 1998. Influenza A virus lacking the NS1 gene replicates in interferon-deficient systems. *Virology* 252: 324.

34. Garcia-Sastre, A. 2001. Inhibition of interferon-mediated antiviral responses by influenza A viruses and other negative-strand RNA viruses. *Virology* 279: 375.

35. Bergmann, M., Garcia-Sastre, A., Carnero, E., Pehamberger, H., Wolff, K., et al. 2000. Influenza virus NS1 protein counteracts PKR-mediated inhibition of replication. *J. Virol.* 74: 6203.

36. Talon, J., Horvath, C. M., Polley, R., Basler, C. F., Muster, T., Palese, P., and Garcia-Sastre, A. 2000. Activation of interferon regulatory factor 3 is inhibited by the influenza A virus NS1 protein. *J. Virol.* 74: 7989.

37. Wang, X., Li, M., Zheng, H., Muster, T., Palese, P., Beg, A. A., and Garcia-Sastre, A. 2000. Influenza A virus NS1 protein prevents activation of NF-κB and induction of alpha/beta interferon. *J. Virol.* 74: 11566.

38. Goodman, A. G., Smith, J. A., Balachandran, S., Perwitasari, O., Proll, S. C., Thomas, M. J., Korth, M. J., Barber, G. N., Schiff, L. A., and Katze, M. G. 2007. The cellular protein P58[IPK] regulates influenza virus mRNA translation and replication through a PKR-mediated mechanism. *J. Virol.* 81: 2221.

39. Yan, W., Frank, C. L., Korth, M. J., Sopher, B. L., Novoa, I., Ron, D., and Katze, M. G. 2002. Control of PERK eIF2alpha kinase activity by the endoplasmic reticulum stress-induced molecular chaperone P58[IPK]. *Proc. Natl. Acad. Sci. USA* 99: 15920.

40. Suhrbier, A., and La Linn, M. 2004. Clinical and pathologic aspects of arthritis due to Ross River virus and other alphaviruses. *Curr. Opin. Rheumatol.* 16: 374.

41. Luers, A. J., Adams, S. D., Smalley, J. V., and Campanella, J. J. 2005. A phylogenomic study of the genus Alphavirus employing whole genome comparison. *Comp. Funct. Genom.* 6: 217.

42. Brisse, S., Iteman, I., and Schuffenecker, I. 2007. Chikungunya outbreaks. *N. Engl. J. Med.* 356: 2650.

43. Soden, M., Vasudevan, H., Roberts, B., Coelen, R., Hamlin, G., Vasudevan, S., and La Brooy, J. 2000. Detection of viral ribonucleic acid and histologic analysis of inflamed synovium in Ross River virus infection. *Arthritis Rheum.* 43: 365.

44. Lidbury, B. A., Simeonovic, C., Maxwell, G. E., Marshall, I. D., and Hapel, A. J. 2000. Macrophage-induced muscle pathology results in morbidity and mortality for Ross River virus-infected mice. *J. Infect. Dis.* 181: 27.

45. Morrison, T. E., Whitmore, A. C., Shabman, R. S., Lidbury, B. A., Mahalingam, S., and Heise, M. T. 2006. Characterization of Ross River virus tropism and virus-induced inflammation in a mouse model of viral arthritis and myositis. *J. Virol.* 80: 737.

46. Linn, M. L., Aaskov, J. G., and Suhrbier, A. 1996. Antibody-dependent enhancement and persistence in macrophages of an arbovirus associated with arthritis. *J. Gen. Virol.* 77: 407.

47. Porterfield, J. S. 1986. Antibody-dependent enhancement of viral infectivity. *Adv. Virus Research* 31: 335.

48. Suhrbier, A., and Linn, M. L. 2003. Suppression of antiviral responses by antibody-dependent enhancement of macrophage infection. *Trends Immunol.* 24: 165.

49. Lidbury, B.A., and Mahalingam, S. 2000. The specific ablation of antiviral gene expression in macrophages by antibody-dependent enhancement of Ross River virus infection. *J. Virol.* 74: 8376.

50. Mahalingam, S., and Lidbury, B. A. 2002. Suppression of lipopolysaccharide induced antiviral transcription factor (STAT-1 and NF-κB) complexes by antibody-dependent enhancement of macrophage infection by Ross River virus. *Proc. Natl. Acad. Sci. USA* 99: 13819.

7 Ethical Considerations for a Genetic Future in Diagnosis and Drug Development

Lexie A. Brans
Faculty of Health, University of Canberra

Brett A. Lidbury
Centre for Biomolecular and Chemical
Sciences, University of Canberra

CONTENTS

7.1   INTRODUCTION

With the availability of the full human genome sequence, the genome sequences of other species, and rapid progress in the development of technologies with which to analyze and apply genetic insights and discoveries, a dominant future for genetic intervention and individual gene profiling in medicine is certain. This book has examined some exciting scientific developments and insights that will contribute to the gene profiling of patients, and the application of genetic knowledge to new drug designs. However, while such scientific progress provides potential solutions to health problems (i.e., more accurate diagnosis of disease and drugs of enhanced efficacy), there are also potential new problems created because of the accuracy and power of gene-based technologies, such as those described in this volume. These additional challenges apply outside of the cellular, molecular, and biochemical details of the genetic revolution, to the impact of this science on individuals and on their societies. Therefore, while accepting the scientific benefits of a given scientific discovery and its subsequent application to human health, it is essential that bioethical perspectives also be captured as part of the incorporation of genetic technologies into the standard practice of health and medicine. Decisions must be made on the application of new genetic technologies, and bioethics provides an intellectual platform from which to make these decisions.

Bioethics is a relatively recent phenomenon in health, arising from the much older disciplines of philosophy and medical ethics. Its aim is to come to a greater understanding about complex issues and to address profound philosophical questions such as what it means to be human [1]. This chapter provides an overview of some of these important considerations, with specific reference to gene profiling and associated drug design. The chapter does not purport to be a definitive analysis of bioethics relative to advances in gene technology and science. Rather, it is more a flavor of the topic to stimulate thought and to provide guidance to scientists beginning their ethical inquiry into the implications of their work in genetics and gene technology.

7.2   BECAUSE WE CAN, WE SHOULD...?

Any scientific breakthrough or technological advance brings with it the capacity to perform tasks that were not previously possible without the new technology, or with an earlier superseded version of the technology. The instinct to benefit by taking advantage of such advances is understandable, but should always be prefaced by the "because we can, we should...?" consideration. This question forces the broader contemplation of what is now possible to achieve technically in terms of the wider impact on individuals or society and an assessment of whether or not the technical possibilities should indeed be acted upon, and to what end. With the advance of science and subsequent technologies that emanate from basic discoveries, we now have the burden of boundless choice concerning treatment and diagnoses, and the need to make challenging decisions about how to implement the technologies. This burden, and the exciting possibilities new technologies open up, is an example of a classic philosophical analysis along the lines of identifying and justifying the means and the end [2].

With such ever-expanding choice comes growing complexity, and with this complexity comes the need for a framework to think through, in a careful and considered way, what might be the implications of the technology, not only for the individual concerned but for society as a whole. For example, who is best placed to make the decisions and what weight should be given to individual decision makers? Also, who decides which decisions are valid, and using what criteria? Should the decisions of the individual concerned be paramount, or are those of the family to take precedence? On the other hand, is it best to defer to "experts," such as medical and legal professionals and/or bioethical experts? To what extent should legislation be developed and enforced to facilitate (desired) behaviors that allegedly benefit the wider society? At heart, these are all ethical issues, and bioethics provides a framework with which to approach the increasing complexity resulting from technological and scientific advances. In this sense, bioethics is a great deal more than the submission of a form for clearance by a research ethics committee, notwithstanding the importance of obtaining that kind of approval prior to a research project or clinical trial.

7.3   APPLIED ETHICS

The use of bioethical approaches as a means of making decisions in the face of complexity and uncertainty, especially in health care, has come to be called applied ethics. As its name suggests, applied ethics attempts to apply philosophical interpretations about ethics to practical professional problems and issues [3–5].

For a professional, the aim of applied ethics is to decide on appropriate moral actions in given situations via the application of ethical principles and theories. This process is also known as normative or action-guiding ethics [5]. The decision-making processes take into account the legitimate moral claims of all those involved in the particular circumstance/s as well as providing justifications for the decisions themselves in order to demonstrate that the decisions extend beyond mere self-interest or preference. Individuals can make these kinds of ethical decisions on their own behalf and act on them alone, for instance, the actions of a scientist in his or her normal day-to-day work. Alternatively, decisions can be made by groups and acted upon by the individuals in the groups, for instance, the actions of a research team in a clinical trial. The decisions made can directly affect individuals in single circumstances, or they may be generalized to larger groups, and to society as a whole.

Society's ethical expectations of particular professional groups may be developed by that profession, and in that case are typically explicated in professional codes of ethics and of conduct [6]. Alternatively, they may be developed and published by governments, for example, the Australian National Health and Medical Research Council's *Australian Code for the Responsible Conduct of Research* [7].

7.3.1   MORALS AND ETHICS

Broad consideration of the meaning and background to the philosophical constructs of ethics and morals is pertinent at this point. The two words have their lexical origins in ancient Greek and Latin, respectively, and most philosophers do not distinguish between them [5]. In popular parlance, however, they are sometimes distinguished as ethics being the ideals we aspire to and morals as being the minimum standards of professional behavior and attitude [6]. To put it even more simply, morals can be seen as the actions and ethics as the theory that informs the actions. In this chapter the two words are used interchangeably.

7.3.2   BIOETHICAL PRINCIPLES AND APPLIED ETHICS

In the Western philosophical tradition, there are many different approaches to bioethics and applied ethics, including feminist epistemology and philosophy [3–5, 8]. Perhaps the most influential of them, however, are the bioethical principles articulated by the authors Beauchamp and Childress in their seminal text, first published in 1979, *Principles of Biomedical Ethics* [9]. The principles include autonomy, beneficence, nonmaleficence, and justice, and they have been enthusiastically embraced and applied by all kinds of professional groups in health care and beyond [5], although the approach is not without its critics [10–12].

The application of these principles is meant to encourage and facilitate critical and careful thinking in order to reach reasoned and justifiable conclusions about actions that can be defended while, at the same time, being practical in their outcomes. The defense is usually in the form of rational arguments and often includes evidence-based science, either in the actual decision-making process or in the justification itself. Since all health care is directed to benefits rather than harms, a justification for an ethical action is always required, although the actual form and content of that justification may vary. Robust justifications are required so that the community can be confident that ethical decisions will be more likely than not to promote best interests rather than simply serve an individual whim, preference, or prejudice.

7.3.3 Autonomy, Beneficence, and Nonmaleficence

Autonomy, beneficence, and nonmaleficence are perhaps the most significant of the principles, since they are at the heart of important concepts in the Western world, namely, the notion of reciprocal rights and duties. Rights are generally thought of as applying to an individual, and where they apply, others have a duty to act to respect and facilitate those rights. Rights may also be accorded to collectives or groups, e.g., as in consumer rights, and once again others have duties to fulfill to honor and respect those rights [9]. When rights clash with duties, as they so often do in a health-care system delivered in a society characterized by ethical pluralism [10], then a classical ethical dilemma arises. It may even be the case that what is to constitute a right, and whether or not a duty is then owed, constitutes the nature of the dilemma itself, e.g., as may occur in randomized controlled trials [13]. There is a vast literature on these kinds of topics in health care, and chapters 1 and 2 of the Beauchamp and Childress text [9] and chapters 3 and 4 of the Preston text [5] are recommended as beginning references for the articulation of moral norms and principles, ideal virtues and moral excellence, and the application of ethical theories in responsible decision making in the context of professional life.

We turn now to a brief description of the salient features of the three principles.

7.3.3.1 Autonomy

Autonomy concerns the right of a person to make, and to then act on, decisions they make about themselves and their bodies. Its essential feature is self-determination and the exercise of free will, not in a capricious sense, but in a sense where one's values, expressed as choices, show a degree of consistency over time [3, 4, 9]. So, for example, a person may make a decision to accept or decline certain treatment, based on what that individual judges to be in his or her best interests. An example is when a person makes decisions about treatments in order to follow the dictates of his or her religious faith [5, 10]. Ideally, autonomous decisions are freely made, are based on adequate information provided in a form that can be readily understood by the person making the decision, and where the decision is made in an atmosphere free from coercion of any kind. In that kind of circumstance, then, generally speaking, an autonomous decision has been made.

When others act on these kinds of considered and autonomous decisions, then they are said to be acting out of respect for autonomy [9]. In so doing, they are fulfilling a duty to protect the rights of individuals to decide what will or will not happen to them, physically and psychologically. That right is protected in law through the legal principle of informed consent [14, 15], although autonomy and informed consent, in the context of genetic testing, has its own set of problems, for example, confidentiality [16].

7.3.3.2 Beneficence

Beneficence is essentially concerned with benefiting, i.e., with acting so that the actions of one person result in an overall balance of benefits to others while at the same time not resulting in undue harm to the individual. Beauchamp and Childress [9] have developed conditions that they regard as necessary to fulfill to determine when, and if, one person owes another an obligation to benefit [9, pp. 170–176]. Of course a health-care professional may experience a dilemma when he or she can foresee that respecting autonomy is likely to result in harm for that person. These dilemmas are generally discussed in terms of paternalism [3, 4, 9], and they can be extremely complex, especially in genetics [16].

7.3.3.3 Nonmaleficence

Nonmaleficence is a closely related principle, since it concerns benefiting (or doing good) through refraining from doing some harmful action. It is often expressed in the form of the maxim, "Above all, do no harm" [9]. When harm to another can reasonably be foreseen, and that harm can be prevented by someone not doing something, e.g., not giving a drug that would constitute an overdose, then there is an obligation to act to benefit, or to do good. The legal principle pertaining to both beneficence and nonmaleficence is negligence [14]. Once again, conflicts between acting to benefit rather than causing harm can create dilemmas, e.g., whether or not the professional should offer his or her specialist knowledge to enhance autonomous decision making, especially if the provision of that information is likely to make a difficult decision-making process even more onerous for the individual concerned.

7.3.3.4 Ethical Dilemmas

An ethical dilemma occurs when we are faced with making a decision where the options available do not clearly indicate that one action has more benefits than another. In other words, there are moral imperatives or duties to act, and in respecting one, we can't at the same time respect the other. On the other hand, if it is clear that one action has more benefits than another, then it is likely to be a clinical practice situation with complex ethical dimensions, rather than a dilemma. It needs to be remembered that refraining from taking an action can be the action; namely, not to act can have as much moral weight as acting (the nonmaleficence principle).

Essentially, then, a dilemma occurs when we know that whatever action we take, there will be both benefits and harms: a "damned if you do and damned if you don't" circumstance. For examples of dilemmas, refer to the *Dax Cowart* case in Widdershoven [11] and the variety of cases discussed in Beauchamp and Childress [9].

In health care, we are aiming for overall benefits, not harms, and where we can reasonably foresee harms arising, then we are obliged to fulfill a duty to strive to limit those harms and maximize the benefits. Where we cannot do both simultaneously, or where we anticipate that more harms are likely whatever the course of action we take, then we probably have an ethical dilemma. It is often the case too that a "gut reaction" about something being "bad" or "wrong" can indicate that an ethical dilemma has arisen, a situation where we need to think through, in a careful and considered fashion, what course of action is the ethical one, and what is the justification for that decision.

The application of ethical principles and the associated process of thinking through the options (the normative-ethics approach noted earlier) not only bring to prominence the essential nature of the situation, they also lead to more robust and defensible decision making. It is at this thinking-through stage that objective evidence may be called upon, for example, to clarify the clinical circumstances of the case. Professional codes of ethics and of conduct, as well as legal principles, may also be drawn upon as part of the decision-making process. When none of these avenues, clearly and unequivocally, indicates the direction to take, then there is an ethical dilemma. In that case, the application of ethical principles (and theories [5]) will bring the most clarity and provide the best guidance in what are frequently emotionally charged and troubling circumstances.

The above description of the decision-making processes involved in an ethical dilemma can sound overly tedious and unnecessary. Professional life is full of ethics, though, and we "do" ethics all the time, even when we are not consciously aware of it. Thinking through each situation in the manner described, either as an intellectual classroom exercise or in the real world of clinical practice, enables the practitioner to not only reach a decision, but to also build up a repertoire of ethical skills that can be drawn upon in subsequent and similar scenarios. In this respect, developing ethical expertise is not dissimilar to building up skills and expertise in a professional discipline. As the Canadian bioethicist and lawyer Margaret Somerville notes, "doing" ethics takes time [17], and it cannot be applied like a "cookie cutter" to any one situation [18]. That is not to say that we "waste" time deliberating while the patient suffers. It is rather that, once developed, then already-learned ethical skills and expertise can be drawn upon when a rapid and robust, yet fully defensible, ethical decision is required. A mentor is an invaluable asset in building up an ethical repertoire, and all readers are encouraged to find their own mentor to further develop this aspect of their professional lives.

7.4 BIOETHICAL PRINCIPLES AND THE NEW GENETICS

To illustrate these issues in the context of "gene profiling in drug design," the following future hypothetical dilemma is suggested, followed by questions that may be asked around this dilemma when applying the three ethical principles being discussed.

A patient from a family with a history of Huntington's disease (HD) is genetically screened for the HD disease gene, located on chromosome 4. With HD causing neurological dysfunction in middle age, the genetic screening also includes specific genetic markers for other brain proteins associated with the HD phenotype. The results come back as negative for the HD allele, but a genetic defect was detected for another key brain protein that will possibly lead to disease within the next five years. While the patient will be told about the expected HD result, should the patient also be informed about the other mutation of likely health consequence?

This hypothetical scenario presents us with a situation where genetic technology has advanced to allow the detection of new gene mutations that were not previously tested, thus creating an ethical dilemma. Normally the patient would have been oblivious to the presence or consequence of this gene and its biological function. Importantly, having the mutated gene may not necessarily lead to a disease phenotype (whereas with a purely genetic disease like HD, an autosomal dominant disorder, an adverse health outcome, is definite). In other words, the new genetic results do not unequivocally indicate that the patient will inevitably progress to a certain disease outcome, as they will with HD. Therefore, we have the ethical dilemma of whether or not to inform the patient of these wider genetic findings as revealed by laboratory testing, i.e., the dilemma between the patient's right to know, or to not know [19], information that may have a significant bearing on the patient's ability to make autonomous decisions about his or her own body. This is a classic tussle between benefiting by providing information in order to facilitate autonomy (a right to know versus a duty to inform), or not providing information in order to prevent harm (a duty not to harm [nonmaleficence] versus a right to make a fully informed autonomous decision).

This dilemma is compounded by questions such as: Do we also have a duty to inform the patient's family, employer, and life insurance company of the findings? If the answer is "yes" for any of these choices, what are the moral reasons for that decision? A further ethically based question is, to what extent do we need to involve the patient in that decision-making process, and for what moral reasons?

Applying the ethical principles of autonomy, beneficence, and nonmaleficence to the above gene-profiling dilemma illustrates the positive role for bioethics in the challenges presented by advances in human genetics. Rather than launching into a comprehensive analysis of the complexities of this situation from the perspectives of all those involved, or providing a detailed exploration of the ethical (and other) ramifications of all the available options, we pose some key questions to address around each principle and suggest how each principle may assist in the necessary ethical deliberations. The readers are invited to reflect on the questions we pose, to add others they consider relevant, and to then reflect on what they would do in this hypothetical scenario, and for what moral reasons.

For the purposes of this exercise, we have assumed the following: that the patient is fully competent to make decisions in both a legal and a clinical sense; that the patient has made autonomous decisions in a noncoercive context, including the decision to have the test in the first place; that there are no reasonable barriers to giving the patient the information, legal or otherwise; and that the laboratory is fully competent in the testing procedures, supported by the appropriate quality control measures, so we can be confident of the veracity of the results.

As an aside, this hypothetical scenario also illustrates the multidisciplinary nature of bioethics, e.g., the roles of lawyers, medical scientists, cultural experts, and genetic counselors. If, in time, this hypothetical scenario became reality, then the expertise of genetic counselors would be called upon, for example, in a formal counseling session with the person (and the person's family). In any case, the ethical principles they would use and apply in assisting the careful thinking-through process will most likely be those we have identified.

We turn now to the considerations to be taken into account in the application of autonomy, beneficence, and nonmaleficence to the above hypothetical scenario. Identify the exact nature of this dilemma by asking the following kinds of questions.

1. *Who are the moral actors (or moral agents) in the case?*

   It is important to identify who they are because, as a moral actor or agent [5], each person has various moral rights to certain things, such as information, and each also has a reciprocal duty to fulfill to others, such as acting on information in order to benefit and not harm. Deciding what is to count as a benefit (beneficence) and what is to count as a harm is, of course, part of the thinking-through process. Objective evidence, such as that provided by randomized controlled trials, is frequently used to assist in making these kinds of decisions, i.e., the evidence will show if option x will benefit. Note that the evidence alone will not necessarily resolve the dilemma; it merely clarifies and so contributes to a robust and defensible decision.

   The moral actors include those who will need to make decisions about whether or not to tell, who will act on those decisions, and who is told or not told. Should the professional provide everyone with the same information and in exactly the same format? To what extent should each of the moral actors be involved in the decision-making processes, namely, to what extent are we obliged to respect autonomy? To what extent do cultural mores impinge on this dilemma? In some cultures, for instance, the family makes the autonomous decision, not the individual alone [7]. From the perspective of the health-care professionals, it is also not always the case that the person who makes the decision is also the one who acts. In this hypothetical scenario, for instance, a team may make a decision to tell (or not tell), but an individual professional may be the one who tells the patient and/or the family.

2. *Who should be told the test results?*

   The key consideration here is to what end that information is being passed on, and the reason for doing so (the "because we can, we should" notion). If autonomy is considered paramount, as it so often is in the Western world, then we tell only the patient. All other decisions about who is told and what they will be told are made by the patient, probably in consultation with the professionals, and perhaps a genetic counselor. It may be the case that the patient decides to involve the family, perhaps for cultural reasons, but essentially, if we are benefiting by respecting autonomy, then we tell only the patient in the first instance. In that case, it has been decided that the patient has a right to know the results of the tests so that they may make subsequent informed and freely chosen decisions, for instance, about whether or not to take steps to ameliorate the effects of the impending brain disease.

If the decision is made not to pass on the information, e.g., on the grounds that the information is unnecessarily burdensome for the patient (the non-maleficence principle), then subsequent questions arise, including rights to know or not to know [19] versus duties to not harm (by withholding information) and issues around paternalism.

3. *Who is best placed to pass on information and how should it be done?*

Since it is likely to be a sensitive and emotionally charged situation, the team member with the most experience in these matters is generally chosen as the one who does the telling. The assumption here is that someone with more experience will be able to maximize benefits (the beneficence principle) and reduce any harms that may arise from well-meaning yet inexperienced professionals providing information (the nonmaleficence principle). With the patient's permission (the autonomy principle), the more junior professional may sit in on the interview in order to expand his or her own experience. In that circumstance, this is an example of developing ethical expertise with a mentor as he or she "does" ethics. It is also an example of beneficence in action, since the more junior professional benefits by learning, and ultimately all patients will benefit from that expanded expertise.

There will also need to be a decision made as to whether or not the patient will be told alone or in the company of family members. Will the patient, who is the principal moral actor, be given the opportunity to make that decision him- or herself (autonomy and beneficence in action), or will there be no need to consult because we already know what the patient wants, i.e., we already know that the patient's autonomous choice is to be told the test information, irrespective of the results?

4. *How and in what form will the information be passed on?*

This is private and sensitive information, so a quiet room where conversations cannot be overheard would be the best choice. It may be appropriate to offer the patient follow-up counseling. Consider how much information the patient can realistically absorb in this one situation. Would it be useful to invite the patient to have another person of his or her choosing to be present? Do we also need to give the patient some kind of printed information with the option for a follow-up appointment for further discussion? Answers to these questions see the beneficence and nonmaleficence principles in action: respectively, acting to benefit by giving real choices that we then honor with follow-up appointments and refraining from intentionally harming (by considering the room, the persons present, the format of the information), proactively offering further care, and so on.

5. *To what extent can this experience be generalized to other situations?*

The answers to this question illustrate the more universal nature of applied ethics. The actions that have been taken in this case will affect not only the patient, but the professionals involved as well as future patients and their loved ones. So, for instance, this experience will act as a guide for professionals, say for making policy recommendations about what kinds of genetic information can be (or should be) passed on to patients, by whom, in what circumstances, and for what purpose. Who else will and can benefit from these kinds of experiences? Do we, for example, have a duty to enlist the media and other health-care professionals to mount a preventive public health-awareness campaign around the new genetic test results associated with HD testing? Making these kinds of decisions about providing, or not providing, information [2, 19] to current and future patients brings all three principles into action.

7.5   CONCLUSION

This chapter has provided a short introduction to the practical possibilities of the application of bioethical principles to the exciting, yet complicated, future world of genetics in human health. It is certainly the case that as science and technology produce ever more undreamt of possibilities for the treatment and prevention of genetic and other health disorders, the companion to this future universe and future generations will be challenging decisions about how to ensure that those advances benefit rather than harm. We also need to think about how much science will challenge the age-old ethical questions such as what it means to be human. These are questions that have long bedeviled philosophy, and now they are becoming increasingly important for science.

We have suggested that bioethics and applied ethics provide the intellectual and philosophical platforms with which to confront these challenges. Applied ethics cannot be "done" in isolation, and we urge all readers to develop their ethical expertise so that it can be brought to bear on the fascinating and complex ethical problems they will undoubtedly face in the future.

REFERENCES

1. Kuhse, H., and Singer, P. 1998. What is bioethics? A historical introduction. In *A companion to bioethics*, eds. H. Kuhse and P. Singer. Malden, Mass.: Blackwell, p. 3.
2. de Melo-Martin, I. 2006. Genetic testing: the appropriate means for a desired goal. *J. Bioethical Inquiry* 3: 167.
3. Craig, E. 2005. *The shorter Routledge encyclopedia of philosophy.* London: Routledge.
4. Audi, R. 1999. *The Cambridge dictionary of philosophy.* 2nd ed. London: Cambridge University Press.
5. Preston, N. 2007. *Understanding ethics.* 3rd ed. Annandale, Australia: Federation Press.
6. Miller, S. 2002. *Model code of ethics principles.* Professional Standards Council, New South Wales Government, Australia.

7. Australian Government. 2007. *Australian Code for the responsible conduct of research.* National Health and Medical Research Council. Available online. URL: http://www. nhmrc.gov.au/publications/synopses/_files/r39.pdf. Accessed February 19, 2008.
8. Sherwin, S. 2002. Towards a feminist ethics of health care. In *Healthcare ethics and human values: an introductory text with readings and case studies,* eds., K. W. M. Fulford, D. L. Dickenson, and T. H. Murray. Malden, Mass.: Blackwell, p. 25.
9. Beauchamp, T. L., and Childress, J. F. 2001. *Principles of biomedical ethics.* 5th ed. Oxford: Oxford University Press.
10. Charlesworth, M. 2005. Don't blame the "bio"—blame the "ethics": varieties of (bio)ethics and the challenge of pluralism. *J. Bioethical Inquiry* 2: 10.
11. Widdershoven, G. M. 2002. Alternatives to principilism: phenomenology, deconstruction, hermeneutics. In *Healthcare ethics and human values: an introductory text with readings and case studies,* eds. K. W. M. Fulford, D. L. Dickenson, and T. H. Murray. Malden, Mass.: Blackwell, p. 41.
12. Gillon, R., ed. 1994. *Principles of health care ethics.* New York: Wiley and Sons.
13. Rothman, D. J. 2004. Research, human: historical aspects. In *Encyclopaedia of bioethics,* ed. S. G. Post. 3rd ed. New York: Macmillan Reference, p. 2316.
14. Kerridge, I., Lowe, M., and McPhee, J. 2005. *Ethics and law for the health professions.* 2nd ed. Annandale, Australia: Federation Press.
15. Faden, R., and Beauchamp, T. 1986. *A history and theory of informed consent.* New York: Oxford University Press.
16. Neil, D., and Craigie, J. 2004. The ethics of pharmacogenomics. *Monash Bioethics Rev.* 23: 9.
17. Somerville, M. 2000. *The ethical canary: science, society and the human spirit.* Camberwell, Australia: Penguin Books.
18. Small, R. 2001. Codes are not enough: what philosophy can contribute to the ethics of educational research. *J. Philosophy Educ.* 35: 387.
19. McDougal, R. 2004. Rethinking the "right not to know." *Monash Bioethics Rev.* 23: 22.

Index